Study Guide/ Student Solutions Manual to accompany

Principles of

THIRD EDITION

A CALCULUS-BASED TEXT   VOLUME 2

Serway & Jewett

John R. Gordon
James Madison University

Ralph McGrew
Broome Community College

Raymond A. Serway
James Madison University

John W. Jewett, Jr.
California State Polytechnic University-Pomona

BROOKS/COLE

THOMSON LEARNING

Australia • Canada • Mexico • Singapore • Spain

United Kingdom • United States

For more information about our products, contact us at:
Thomson Learning Academic Resource Center
1-800-423-0563

For permission to use material from this text, contact us by:
Phone: 1-800-730-2214
Fax: 1-800-731-2215
Web: www.thomsonrights.com

Asia
Thomson Learning
60 Albert Complex, #15-01
Alpert Complex
Singapore 189969

Australia
Nelson Thomson Learning
102 Dodds Street
South Street
South Melbourne, Victoria 3205
Australia

Canada
Nelson Thomson Learning
1120 Birchmount Road
Toronto, Ontario M1K 5G4
Canada

Europe/Middle East/South Africa
Thomson Learning
Berkshire House
168-173 High Holborn
London WC1 V7AA
United Kingdom

Latin America
Thomson Learning
Seneca, 53
Colonia Polanco
11560 Mexico D.F.
Mexico

Spain
Paraninfo Thomson Learning
Calle/Magallanes, 25
28015 Madrid, Spain

Preface

This <u>Student Solutions Manual and Study Guide</u> has been written to accompany the textbook **Principles of Physics**, Third Edition, Volume II, by Raymond A. Serway and John Jewett. The purpose of this <u>Student Solutions Manual and Study Guide</u> is to provide students with a convenient review of the basic concepts and applications presented in the textbook, together with solutions to selected end-of-chapter problems. This is not an attempt to rewrite the textbook in a condensed fashion. Rather, emphasis is placed upon clarifying typical troublesome points, and providing further practice in methods of problem solving.

Each chapter is divided into several parts, and every textbook chapter has a matching chapter in this book. Very often, reference is made to specific equations or figures in the textbook. Each feature of this Study Guide has been included to insure that it serves as a useful supplement to the textbook. Most chapters contain the following components:

- **Notes From Selected Chapter Sections:** This is a summary of important concepts, newly defined physical quantities, and rules governing their behavior.

- **Equations and Concepts:** This section highlights selected equations that relate to specific concepts developed in each chapter. You should practice expressing each equation in a clear and concise prose statement.

- **Suggestions, Skills, and Strategies:** This section offers hints and strategies for solving typical problems that the student will often encounter in the course. In some sections, suggestions are made concerning mathematical skills that are necessary in the analysis of problems. Note the Pitfall Prevention boxes in the margins of each chapter of the text; careful attention to these concise explanations of common misconceptions will increase your understanding of important concepts.

- **Review Checklist:** This is a list of topics and techniques the student should master after reading the chapter and working the assigned problems.

- **Answers to Selected Conceptual Questions:** Suggested responses are provided for twenty percent of the Conceptual Questions.

- **Solutions to Selected End-of-Chapter Problems:** Solutions are given for approximately half of the odd-numbered problems from the text. Problems were selected to illustrate important concepts in each chapter.

- **Tables:** A list of selected Physical Constants is printed on the inside front cover; and a table of some Conversion Factors is provided on the inside back cover.

A note concerning significant figures: In nearly all problem statements, data is given to three significant figures. The answers to end-of-chapter problems are stated to three significant figures, though intermediate calculation steps are carried out with as many digits as possible. The last digit is uncertain, often depending on the precision of the values assumed for physical constants and properties.

We sincerely hope that this <u>Student Solutions Manual and Study Guide</u> will be useful to you in reviewing the material presented in the text, and in improving your ability to solve problems and score well on exams. We welcome any comments or suggestions which could help improve the content of this study guide in future editions; and we wish you success as you begin exploring the exciting world of Physics.

John R. Gordon
Harrisonburg, VA

Ralph McGrew
Binghamton, NY

John W. Jewett, Jr.
Pomona, CA

Raymond A. Serway
Leesburg, VA

Acknowledgments

It is a pleasure to acknowledge the excellent work of the staff at DSC Publishing, including Mike Rudmin, Jonas Šiaulys, Alexandras Urbonas, and Joseph Rudmin, whose attention to detail in the preparation of the camera-ready copy did much to enhance the quality of this edition of the <u>Student Solutions Manual and Study Guide</u>. Their graphics skills and technical expertise combined to produce illustrations for the first edition which continue to add much to the appearance and usefulness of this volume.

We appreciate the support given this project by the professional staff of Brooks/Cole Thomson Learning. Our special thanks go to Susan Dust Pashos, Jay Campbell, Ed Dodd, and Charlene Squib. Finally, we express our appreciation to our families for their inspiration, patience, and encouragement.

Suggestions for Study

Very often we are asked "How should I study physics and prepare for examinations?" There is no simple answer to this question, however, we would like to offer some suggestions which may be useful to you.

1. It is essential that you understand the basic concepts and principles before attempting to solve assigned problems. This is best accomplished through a careful reading of the textbook before attending your lecture on that material, jotting down certain points which are not clear to you, taking careful notes in class, and asking questions. You should reduce memorization of material to a minimum. Memorizing sections of a text, equations, and derivations does not necessarily mean you understand the material. Perhaps the best test of your understanding of the material will be your ability to solve the problems in the text, or those given on exams.

2. Try to solve as many problems at the end of the chapter as possible. You will be able to check the accuracy of your calculations to the odd-numbered problems, since the answers to these are given at the back of the text. Furthermore, detailed solutions to approximately half of the odd-numbered problems are provided in this study guide. Many of the worked examples in the text will serve as a basis for your study.

3. The method of solving problems should be carefully planned. First, read the problem several times until you are confident you understand what is being asked. Look for key words which will help simplify the problem, and perhaps allow you to make certain assumptions. You should also pay special attention to the information provided in the problem. In many cases a simple diagram is a good starting point; and it is always a good idea to write down the given information before trying to solve the problem. After you have decided on the method you feel is appropriate for the problem, proceed with your solution. If you are having difficulty in working problems, we suggest that you again read the text and your lecture notes. It may take several readings before you are ready to solve certain problems, though the solved problems in this Study Guide should be of value to you in this regard. However, your solution to a problem does not have to look just like the one presented here. Remember, different approaches can often lead to the same correct solution to a problem. If you wonder about the validity of an alternative approach, ask your instructor.

4. After reading a chapter, you should be able to define any new quantities that were introduced, and discuss the first principles that were used to derive fundamental formulas. A review is provided in each chapter of the Study Guide for this purpose, and the marginal notes in the textbook (or the index) will help you locate these topics. You should be able to correctly associate with each physical quantity the symbol used to represent that quantity (including vector notation, if appropriate) and the SI unit in which the quantity is specified. Furthermore, you should be able to express each important formula or equation in a concise and accurate prose statement.

5. We suggest that you use this Study Guide to review the material covered in the text, and as a guide in preparing for exams. You should also use the Chapter Review, Notes From Selected Chapter Sections, and Equations and Concepts to focus in on any points which require further study. Remember that the main purpose of this Study Guide is to improve upon the efficiency and effectiveness of your study hours and your overall understanding of physical concepts. However, it should not be regarded as a substitute for your textbook or individual study and practice in problem solving.

Table of Contents

Chapter 16

Temperature and the Kinetic Theory of Gases

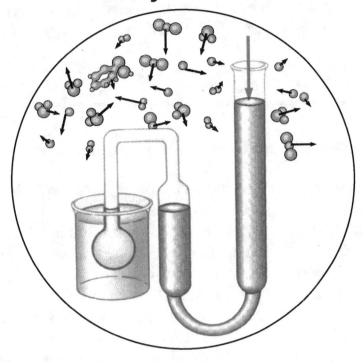

INTRODUCTION

A quantitative description of thermal phenomena requires a careful definition of the concepts of temperature, heat, and internal energy. The laws of thermodynamics provide us with relationships among heat, work, and the internal energy of a system.

The composition and structure of a body are important factors when dealing with thermal phenomena. For example, liquids and solids expand only slightly when heated, whereas gases expand appreciably when heated. If the gas is not free to expand, its pressure rises when heated. Certain substances may melt, boil, burn, or explode.

This chapter concludes with a study of ideal gases. We shall approach this study on two levels. The first will examine ideal gases on the macroscopic scale. Here we shall be concerned with the relationships among such quantities as pressure, volume, and temperature. On the second level, we shall examine gases on a microscopic scale, using a model that pictures the components of a gas as small particles. The latter approach, called the **kinetic theory** of gases, will help us to understand what happens on the atomic level to affect such macroscopic properties as pressure and temperature.

In the model of kinetic theory, gas molecules move about in a random fashion, colliding with the walls of their container and with each other. Perhaps the most important consequence of this theory is that it shows the relationship between the kinetic energy of molecular motion and the internal energy of the system. Furthermore, the kinetic theory provides us with a physical basis upon which the concept of temperature can be understood.

NOTES FROM SELECTED CHAPTER SECTIONS

16.1 Temperature and the Zeroth Law of Thermodynamics

Thermal physics is the study of the behavior of solids, liquids and gases, using the concepts of heat and temperature. Two approaches are commonly used in this area of science. The first is a **macroscopic** approach, called **thermodynamics,** in which one explains the bulk thermal properties of matter. The second is a **microscopic** approach, called **statistical mechanics,** in which properties of matter are explained on an atomic scale. Both approaches require that you understand some basic concepts, such as the concepts of temperature and heat. As we will see, **all** thermal phenomena are manifestations of the laws of mechanics as we have learned them. For example, internal energy is actually a consequence of the random motions of a large number of particles making up the system.

The concept of the **temperature** of a system can be understood in connection with a measurement, such as the reading of a thermometer. Temperature, a scalar quantity, is a property that can only be defined when the system is in thermal equilibrium with another system. Thermal equilibrium implies that two (or more) systems are at the same temperature.

The **zeroth law of thermodynamics** states that if two systems are in thermal equilibrium with a third system, they must be in thermal equilibrium with each other. The third system can be a calibrated thermometer whose reading determines whether or not the systems are in thermal equilibrium. There are several types of thermometers that can be used.

16.2 Thermometers and Temperature Scales

Thermometers are devices used to define and measure the temperature of a system. All thermometers make use of a change in some physical property with temperature. Some of the physical properties used are (1) the change in volume of a liquid, (2) the change in length of a solid, (3) the change in pressure of a gas held at constant volume, (4) the change in volume of a gas held at constant pressure, (5) the change in electric resistance of a conductor, and (6) the change in color of a very hot object. For a given substance, a temperature scale can be established based on any one of these physical quantities.

The **gas thermometer** is a standard device for defining temperature. In the constant-volume gas thermometer, a low-density gas is placed in a flask, and its volume is kept constant while it is heated or cooled. The pressure is measured as the temperature of the gas is changed. Experimentally, one finds that the temperature is proportional to the absolute pressure.

The **thermodynamic temperature scale** is based on a scale for which the reference temperature is taken to be the **triple point of water**; that is, the temperature and pressure at which water, water vapor, and ice coexist in equilibrium. On this scale, the SI unit of temperature is the **kelvin**, defined as the fraction $1/273.16$ of the temperature of the triple point of water.

16.5 The Kinetic Theory of Gases

A microscopic **model of an ideal gas** is based on the following assumptions:

The number of molecules is large, and the average separation between them is large compared with their dimensions. Therefore, the molecules occupy a negligible volume compared with the volume of the container.

The molecules obey Newton's laws of motion, but the individual molecules move in a random fashion. By random fashion, we mean that the molecules move in all directions with equal probability and with various speeds. This distribution of velocities does not change in time, despite the collisions between molecules.

The molecules undergo elastic collisions with each other. Thus, the molecules are considered to be structureless, and for any pair of colliding molecules as a system both kinetic energy and momentum are conserved. **The forces between molecules are negligible except during a collision.** The forces between molecules are short-range, so that the only time the molecules interact with each other is during a collision.

The gas is in thermal equilibrium with the walls of the container. The collisions of molecules with the walls are perfectly elastic.

The gas under consideration is a pure gas. That is, all molecules are identical.

EQUATIONS AND CONCEPTS

Pressure is force per unit area and has units of N/m^2. The SI unit of pressure is the pascal (Pa).

$$1\,Pa = 1\,N/m^2$$

One atmosphere of pressure is atmospheric pressure at sea level.

$$1\,atm = 1.013 \times 10^5\,Pa$$

The **Celsius temperature** T_C is related to the absolute temperature T (in kelvins) according to Equation 16.1, where 0 °C corresponds to 273.15 K.

$$T_C = T - 273.15 \qquad (16.1)$$

The **Fahrenheit temperature** T_F can be converted to degrees Celsius using Equation 16.2. Note that 0 °C = 32 °F and 100 °C = 212 °F.

$$T_F = \frac{9}{5}T_C + 32\ {}^\circ F \qquad (16.2)$$

If a body has a length L_i, the **change in its length** ΔL due to a change in temperature is proportional to the change in temperature and the original length. The proportionality constant α is called the **average coefficient of linear expansion.**

$$\Delta L = \alpha L_i \Delta T \qquad (16.4)$$

or

$$\alpha = \frac{1}{L_i}\frac{\Delta L}{\Delta T}$$

If the temperature of a body of volume V_i changes by an amount ΔT at constant pressure, **the change in its volume** is proportional to ΔT and the original volume. The constant of proportionality β is the **average coefficient of volume expansion.** For an isotropic solid, $\beta \cong 3\alpha$.

$$\Delta V = V_f - V_i = \beta_i V \Delta T \qquad (16.6)$$

The number of moles in a sample of any substance equals the ratio of the mass of the sample to the molar mass characteristic of that particular substance.

$$n = \frac{m_{substance}}{M} \qquad (16.8)$$

If n moles of a dilute gas occupy a volume V, the **equation of state**, which relates the variables P, V, and T at equilibrium is that of an **ideal gas**, Equation 16.9, where R is the **universal gas constant.**

$$PV = nRT \qquad (16.9)$$

$$R = 8.315 \, \text{J} / \text{mol·K} \qquad (16.10)$$

or $R = 0.0821 \, \text{L} \cdot \text{atm} / \text{mol} \cdot \text{K}$

The **equation of state of an ideal gas** can also be expressed in the form of Equation 16.11, where N is the total number of gas molecules and k_B is **Boltzmann's constant.**

$$PV = Nk_BT \qquad (16.11)$$

$$k_B = \frac{R}{N_A} = 1.38 \times 10^{-23} \, \text{J} / \text{K} \qquad (16.12)$$

In the kinetic theory of an ideal gas, one finds that the pressure of the gas is proportional to the number of molecules per unit volume and the average translational kinetic energy per molecule. Note that m here represents the mass of a single molecule.

$$P = \frac{2}{3}\left(\frac{N}{V}\right)\left(\frac{1}{2}m\overline{v^2}\right) \qquad (16.13)$$

From Equation 16.13, and the equation of state for an ideal gas, $PV = Nk_BT$, we find that **the absolute temperature of an ideal gas is a direct measure of the average molecular kinetic energy.**

$$\frac{1}{2}m\overline{v^2} = \frac{3}{2}k_BT \qquad (16.15)$$

The **total translational kinetic energy** E of N molecules (or n moles) of a monatomic ideal gas is proportional to the absolute temperature.

$$E_{total} = N\left(\frac{1}{2}m\overline{v^2}\right) = \frac{3}{2}Nk_BT = \frac{3}{2}nRT \qquad (16.17)$$

The expression for the root-mean-square speed of molecules shows that at a given temperature, lighter molecules move faster on the average than heavier molecules.

$$v_{rms} = \sqrt{\overline{v^2}} = \sqrt{\frac{3k_BT}{m}} = \sqrt{\frac{3RT}{M}} \qquad (16.19)$$

REVIEW CHECKLIST

▷ Understand the concepts of thermal equilibrium and thermal contact between two bodies, and state the zeroth law of thermodynamics.

▷ Discuss some physical properties of substances, which change with temperature, and the manner in which these properties are used to construct thermometers.

▷ Describe the operation of the constant-volume gas thermometer and how it is used to define the ideal gas temperature scale.

▷ Convert between the various temperature scales, especially the conversion from degrees Celsius into kelvins, degrees Fahrenheit into kelvins, and degrees Celsius into degrees Fahrenheit.

▷ Provide a qualitative description of the origin of thermal expansion of solids and liquids; define the linear expansion coefficient and volume expansion coefficient for an isotropic solid, and learn how to deal with these coefficients in practical situations involving expansion or contraction.

▷ State and understand the assumptions made in developing the molecular model of an ideal gas.

▷ Recognize that the temperature of an ideal gas is proportional to the average molecular kinetic energy. State the theorem of equipartition of energy, noting that each degree of freedom of a molecule contributes an equal amount of energy, of magnitude $\frac{1}{2}k_BT$.

ANSWERS TO SELECTED CONCEPTUAL QUESTIONS

1.   A piece of copper is dropped into a beaker of water. If the water's temperature rises, what happens to the temperature of the copper? Under what conditions are the water and copper in thermal equilibrium?

Answer   If the water's temperature increases, that means that energy is being transferred by heat to the water. This can happen either if the copper changes phase from liquid to solid, or if the temperature of the copper is above that of the water, and falling.

In this case, the copper is referred to as a "piece" of copper, so it is already in the solid phase; therefore, its temperature must be falling. When the temperature of the copper reaches that of the water, the copper and water will reach equilibrium, and the subsequent net energy transfer by heat between the two will be zero.

□   □   □   □

6.   If a helium-filled balloon initially at room temperature is placed in a freezer, will its volume increase, decrease, or remain the same?

Answer   The helium is far from liquefaction. Therefore, we model it as an ideal gas, described by $PV = nRT$. In this case, the pressure stays nearly constant, being equal to 1 atm. Since the temperature may decrease by 10% (from 293 K to 263 K), the volume should also **decrease** by 10%. This process is called "isobaric cooling", or "isobaric contraction."

□   □   □   □

13.   An alcohol rub is sometimes used to help lower the temperature of a sick patient. Why does this help?

Answer   As the alcohol evaporates, high-speed molecules leave the liquid. This reduces the average speed of the remaining molecules. Since the average speed is lowered, the temperature of the alcohol is reduced. This process helps to carry energy away from the skin of the patient, resulting in cooling of the skin. The alcohol plays the same role of evaporative cooling as does perspiration, but alcohol evaporates much more quickly than perspiration.

SOLUTIONS TO SELECTED END-OF-CHAPTER PROBLEMS

1.   A constant-volume gas thermometer is calibrated in dry ice (i.e., carbon dioxide in the solid state, which has a temperature of –80.0 °C) and in boiling ethyl alcohol (78.0 °C). The two pressures are 0.900 atm and 1.635 atm. (a) What Celsius value of absolute zero does the calibration yield? What is the pressure at (b) the freezing point of water and (c) the boiling point of water?

Solution

Since we have a linear graph, the pressure is related to the temperature as $P = A + BT$, where A and B are constants.

To find A and B, we use the given data: $0.900 \text{ atm} = A + (-80.0 \text{ °C})B$

and $1.635 \text{ atm} = A + (78.0 \text{ °C})B$

Solving these simultaneously, we find $A = 1.272 \text{ atm}$

and $B = 4.652 \times 10^{-3} \text{ atm/°C}$

Therefore, $P = 1.272 \text{ atm} + (4.652 \times 10^{-3} \text{ atm/°C})T$

(a)   At absolute zero, $P = 0 = 1.272 \text{ atm} + (4.652 \times 10^{-3} \text{ atm/°C})T$

   which gives $T = -274 \text{ °C}$ ◊

(b)   At the freezing point of water, $P = 1.272 \text{ atm} + 0 = 1.27 \text{ atm}$ ◊

(c)   And at the boiling point,

   $P = 1.272 \text{ atm} + (4.652 \times 10^{-3} \text{ atm/°C})(100 \text{ °C}) = 1.74 \text{ atm}$ ◊

3. Liquid nitrogen has a boiling point of –195.81 °C at atmospheric pressure. Express this temperature (a) in degrees Fahrenheit, and (b) in kelvins.

Solution

(a) By Equation 16.2, $\qquad T_F = \frac{9}{5}T_C + 32.0\ °F = \frac{9}{5}(-195.81) + 32.0 = -320\ °F$ ◊

(b) Applying Equation 16.1, $\qquad 273.15\ K - 195.81\ K = 77.3\ K$ ◊

Related Comment   A convenient way to remember Equations 16.1 and 16.2 is to remember the freezing and boiling points of water, in each form:

$$T_{freeze} = 32.0\ °F\ = 0\ °C = 273.15\ K$$

$$T_{boil} = 212\ °F\ = 100\ °C$$

To convert from Fahrenheit to Celsius, subtract 32 (the freezing point), and then adjust the scale by the liquid range of the water.

$$\text{Scale} = \frac{(100-0)\ °C}{(212-32)\ °F} = \frac{5\ °C}{9\ °F}$$

A kelvin is the same size change as a degree Celsius, but the kelvin scale takes its zero point at **absolute zero**, instead of the freezing point of water. Therefore, to convert from kelvin to Celsius, subtract 273.15 K.

---

9. The active element of a certain laser is made of a glass rod 30.0 cm long by 1.50 cm in diameter. If the temperature of the rod increases by 65.0 °C, what is the increase in (a) its length, (b) its diameter, and (c) its volume? Assume that the average coefficient of linear expansion of the glass is $9.00 \times 10^{-6}\ °C^{-1}$.

Solution

(a) $\Delta L = \alpha L_i \Delta T = \left(9.00 \times 10^{-6} \ \degree C^{-1}\right)(0.300 \ \text{m})(65.0 \ \degree C) = 1.76 \times 10^{-4} \ \text{m}$ ◊

(b) The diameter is a linear dimension, so the same equation applies:

$\Delta D = \alpha D_i \Delta T = \left(9.00 \times 10^{-6} \ \degree C^{-1}\right)(0.0150 \ \text{m})(65.0 \ \degree C) = 8.78 \times 10^{-6} \ \text{m}$ ◊

(c) The original volume

$$V = \pi r^2 L = \frac{\pi}{4}(0.0150 \ \text{m})^2(0.300 \ \text{m}) = 5.30 \times 10^{-5} \ \text{m}^3$$

Using the volumetric coefficient of expansion, β,

$$\Delta V = \beta V_i \Delta T \cong 3\alpha V_i \Delta T$$

$$\Delta V \cong 3\left(9.00 \times 10^{-6} \ \degree C^{-1}\right)\left(5.30 \times 10^{-5} \ \text{m}^3\right)(65.0 \ \degree C) = 93.0 \times 10^{-9} \ \text{m}^3 \qquad ◊$$

Related Calculation
The above calculation ignores ΔL^2 and ΔL^3 terms. Calculate the change in volume exactly, and compare your answer with the approximate solution above.

The volume will increase by a factor of $\quad \dfrac{\Delta V}{V_i} = \left(1 + \dfrac{\Delta D}{D_i}\right)^2 \left(1 + \dfrac{\Delta L}{L_i}\right) - 1$

$$\frac{\Delta V}{V_i} = \left(1 + \frac{8.78 \times 10^{-6} \ \text{m}}{0.015 \ \text{m}}\right)^2 \left(1 + \frac{1.76 \times 10^{-4} \ \text{m}}{0.300 \ \text{m}}\right) - 1 = 1.76 \times 10^{-3}$$

So, $\quad \Delta V = \dfrac{\Delta V}{V_i} V_i$

$$\Delta V = \left(1.76 \times 10^{-3}\right)\left(5.30 \times 10^{-5} \ \text{m}^3\right) = 93.1 \times 10^{-9} \ \text{m}^3 = 93.1 \ \text{mm}^3 \qquad ◊$$

The answer is virtually identical; the approximation $\beta \cong 3\alpha$ is a good one.

13. A hollow aluminum cylinder 20.0 cm deep has an internal capacity of 2.000 L at 20.0 °C. It is completely filled with turpentine, and then slowly warmed to 80.0 °C. (a) How much turpentine overflows? (b) If the cylinder is then cooled back to 20.0 °C, how far below the cylinder's rim does the turpentine's surface recede?

Solution

(a) When the temperature is increased from 20.0 °C to 80.0 °C, both the cylinder and the turpentine increase in volume by $\Delta V = \beta V \Delta T$.

The overflow,
$$V_{over} = \Delta V_{turp} - \Delta V_{Al}$$

$$V_{over} = (\beta V_i \Delta T)_{turp} - (\beta V_i \Delta T)_{Al} = V_i \Delta \beta \left(_{turp} \alpha\, 3 \,_{Al} \right)$$

$$V_{over} = (2.000 \text{ L})(60.0 \text{ °C})\left(9.00 \times 10^{-4} \text{ °C}^{-1} - 0.720 \times 10^{-4} \text{ °C}^{-1}\right)$$

$$V_{over} = 0.0994 \text{ L} \qquad\qquad \Diamond$$

(b) After warming the whole volume of the turpentine is

$$V' = 2000 \text{ cm}^3 + \left(9.00 \times 10^{-4} \text{ °C}^{-1}\right)\left(2000 \text{ cm}^3\right)(60.0 \text{ °C}) = 2108 \text{ cm}^3$$

The fraction lost is $\dfrac{99.4 \text{ cm}^3}{2108 \text{ cm}^3} = 4.71 \times 10^{-2}$

This also is the fraction of the cylinder that will be empty after cooling:

$$\Delta h = \left(4.71 \times 10^{-2}\right)(20.0 \text{ cm}) = 0.943 \text{ cm} \qquad\qquad \Diamond$$

17. A school auditorium has dimensions 10.0 m × 20.0 m × 30.0 m. How many molecules of air fill the auditorium at 20.0 °C and a pressure of 101 kPa?

Solution   The air is far from liquefaction, so we can model it as an ideal gas: $PV = nRT$

$$n = \frac{PV}{RT} = \frac{\left(1.01 \times 10^5 \text{ N/m}^2\right)\left[(10.0 \text{ m})(20.0 \text{ m})(30.0 \text{ m})\right]}{(8.315 \text{ J/mol} \cdot \text{K})(293 \text{ K})} = 2.49 \times 10^5 \text{ mol}$$

$$N = nN_A = \left(2.49 \times 10^5 \text{ mol}\right)\left(6.022 \times 10^{23} \text{ molecules/mol}\right) = 1.50 \times 10^{29} \text{ molecules} \qquad \Diamond$$

---

19. The mass of a hot-air balloon and its cargo (not including the air inside) is 200 kg. The air outside is at 10.0 °C and 101 kPa. The volume of the balloon is 400 m³. To what temperature must the air in the balloon be heated before the balloon will lift off? (Air density at 10.0 °C is 1.25 kg/m³.)

Solution

The air displaced by the balloon has a volume of 400 m³ at 10.0 °C, and a mass of

$$m_d = \rho V = \left(1.25 \text{ kg/m}^3\right)\left(400 \text{ m}^3\right) = 500 \text{ kg}$$

The weight of the displaced air is $F_{gd} = m_d g = (500 \text{ kg})\left(9.80 \text{ m/s}^2\right) = 4900 \text{ N}$

The weight of the balloon and cargo is $F_{gb} = m_b g = (200 \text{ kg})\left(9.80 \text{ m/s}^2\right) = 1960 \text{ N}$

The weight of the displaced air (the buoyant force) must equal the weight of the balloon and the air inside the balloon, F_{ga}.

Therefore, $F_{gd} = F_{gb} + m_a g$: $m_a = \dfrac{F_{gd} - F_{gb}}{g} = \dfrac{4900 \text{ N} - 1960 \text{ N}}{9.80 \text{ m/s}^2} = 300 \text{ kg}$

We estimate dry air to be 20% O_2 and 80% N_2, with a little Argon. We calculate the air's average molar mass based on the O_2 and N_2, and round it up to 29.0 g/mol to account for the Argon; we then calculate the number of moles of air that are present.

$$n = \frac{m_a}{M_a} = \frac{m_a}{0.80(28.0 \text{ g/mol}) + 0.20(32.0 \text{ g/mol})}$$

$$n = \frac{(300 \text{ kg})(10^3 \text{ g/kg})}{(29.0 \text{ g/mol})} = 1.03 \times 10^4 \text{ mol}$$

Applying the ideal gas law, and solving for the temperature, $T = PV/nR$:

$$T = \frac{(1.01 \times 10^5 \text{ N/m}^2)(400 \text{ m}^3)}{(1.03 \times 10^4 \text{ mol})(8.315 \text{ J/mol·K})} = 470 \text{ K} \qquad \Diamond$$

23. The pressure gauge on a tank registers the gauge pressure, which is the difference between the interior and exterior pressure. When the tank is filled with oxygen (O_2), it contains 12.0 kg of the gas at a gauge pressure of 40.0 atm. Determine the mass of oxygen that has been withdrawn from the tank when the pressure reading is 25.0 atm. Assume the temperature of the tank remains constant.

Solution The ideal gas law states $PV = nRT$

At constant volume and temperature, $\dfrac{P}{n} = \text{constant}$ or $\dfrac{P_1}{n_1} = \dfrac{P_2}{n_2}$

and $n_2 = \left(\dfrac{P_2}{P_1}\right) n_1$

However, n is proportional to m, so $m_2 = \left(\dfrac{P_2}{P_1}\right) m_1 = \left(\dfrac{26.0 \text{ atm}}{41.0 \text{ atm}}\right)(12.0 \text{ kg}) = 7.61 \text{ kg}$

The mass removed is $\Delta m = 12.0 \text{ kg} - 7.61 \text{ kg} = 4.39 \text{ kg}$ $\Diamond$

29. (a) How many atoms of helium gas fill a balloon of diameter 30.0 cm at 20.0 °C and 1.00 atm? (b) What is the average kinetic energy of the helium atoms? (c) What is the rms speed of the helium atoms?

Solution

(a) The volume is

$$V = \frac{4}{3}\pi r^3 = \frac{4}{3}\pi(0.150 \text{ m})^3 = 1.41 \times 10^{-2} \text{ m}^3$$

Now $PV = nRT$

$$n = \frac{PV}{RT} = \frac{(1.013 \times 10^5 \text{ N/m}^2)(1.41 \times 10^{-2} \text{ m}^3)}{(8.315 \text{ N} \cdot \text{m/mol} \cdot \text{K})(293 \text{ K})}$$

$$n = 0.588 \text{ mol}$$

$$N = nN_A = (0.588 \text{ mol})(6.02 \times 10^{23} \text{ molecules/mol})$$

$$N = 3.54 \times 10^{23} \text{ helium atoms} \qquad \Diamond$$

(b) The kinetic energy is

$$\overline{K} = \frac{1}{2}m\overline{v^2} = \frac{3}{2}k_B T$$

$$\overline{K} = \frac{3}{2}(1.38 \times 10^{-23} \text{ J/K})(293 \text{ K}) = 6.07 \times 10^{-21} \text{ J} \qquad \Diamond$$

(c) An He atom has mass

$$m = \frac{M}{N_A} = \frac{4.0026 \text{ g/mol}}{6.02 \times 10^{-23} \text{ molecules/mol}}$$

$$m = 6.65 \times 10^{-24} \text{ g} = 6.65 \times 10^{-27} \text{ kg}$$

So the kinetic energy is

$$\frac{1}{2}(6.65 \times 10^{-27} \text{ kg})\overline{v^2} = 6.07 \times 10^{-21} \text{ J}$$

and

$$v_{rms} = \sqrt{\overline{v^2}} = 1.35 \text{ km/s} \qquad \Diamond$$

31. A cylinder contains a mixture of helium and argon gas in equilibrium at 150 °C. (a) What is the average kinetic energy of each type of gas molecule? (b) What is the rms speed of each type of molecule?

Solution

(a) Both kinds of molecules have the same average kinetic energy.

$$\tfrac{1}{2}m\overline{v^2} = \tfrac{3}{2}k_B T = \tfrac{3}{2}\left(1.38 \times 10^{-23} \text{ J/K}\right)(273+150) \text{ K}$$

$$\overline{K} = 8.76 \times 10^{-21} \text{ J} \qquad\qquad \diamond$$

(b) The root-mean square velocity can be calculated from the kinetic energy:

$$v_{rms} = \sqrt{\overline{v^2}} = \sqrt{\frac{2\overline{K}}{m}}$$

These two gases are noble, and therefore monatomic. The masses of the molecules are:

$$m_{He} = \frac{(4.00 \text{ g/mol})\left(10^{-3} \text{ kg/g}\right)}{6.02 \times 10^{23} \text{ atoms/mol}} = 6.64 \times 10^{-27} \text{ kg}$$

$$m_{Ar} = \frac{(39.9 \text{ g/mol})\left(10^{-3} \text{ kg/g}\right)}{6.02 \times 10^{23} \text{ atoms/mol}} = 6.63 \times 10^{-26} \text{ kg}$$

Substituting these values,

$$v_{rms, \text{ He}} = \sqrt{\frac{2\left(8.76 \times 10^{-21} \text{ J}\right)}{6.64 \times 10^{-27} \text{ kg}}} = 1.62 \times 10^{3} \text{ m/s}$$

and $\quad v_{rms, \text{ Ar}} = \sqrt{\dfrac{2\left(8.76 \times 10^{-21} \text{ J}\right)}{6.63 \times 10^{-26} \text{ kg}}} = 514 \text{ m/s} \qquad\qquad \diamond$

---

35. From the Maxwell-Boltzmann speed distribution, show that the most probable speed of a gas molecule is given by Equation 16.23. Note that the most probable speed corresponds to the point at which the slope of the speed distribution curve, dN_v/dv, is zero.

Solution

From the Maxwell speed distribution function,

$$N_v = 4\pi N \left(\frac{m}{2\pi k_B T}\right)^{3/2} v^2 e^{\left(-mv^2/2k_B T\right)}$$

we locate the peak in the graph of N_v versus v by evaluating dN_v/dv and setting it equal to zero, to solve for the most probable speed:

$$\frac{dN_v}{dv} = 4\pi N \left(\frac{m}{2\pi k_B T}\right)^{3/2} \left(\frac{-2mv}{2k_B T}\right) v^2 e^{\left(-mv^2/2k_B T\right)} + 4\pi N \left(\frac{m}{2\pi k_B T}\right)^{3/2} 2v e^{\left(-mv^2/2k_B T\right)} = 0$$

This equation has solutions $v = 0$ and $v \to \infty$, but those correspond to minimum-probability speeds, so we divide by v and by the exponential function.

$$v_{mp}\left(-\frac{m(2v_{mp})}{2k_B T}\right) + 2 = 0$$

$$v_{mp} = \sqrt{\frac{2k_B T}{m}}$$

This is Equation 16.23.   ◊

_____

43. A liquid has a density ρ. (a) Show that the fractional change in density for a change in temperature ΔT is $\Delta \rho / \rho = -\beta \Delta T$. What does the negative sign signify? (b) Fresh water has a maximum density of 1.0000 g/cm³ at 4.0 °C. At 10.0 °C, its density is 0.9997 g/cm³. What is β for water over this temperature interval?

Solution

We start with the two equations:

$$\rho = \frac{m}{V} \qquad \text{and} \qquad \frac{\Delta V}{V} = \beta \Delta T$$

(a) Differentiating the first equation,

$$dp = -\frac{m}{V^2}\, dV$$

For very small changes in V and ρ, this can be written as

$$\Delta \rho = -\frac{m}{V}\frac{\Delta V}{V} = -\left(\frac{m}{V}\right)\frac{\Delta V}{V}$$

Substituting both of our initial equations, we find that

$$\Delta \rho = -\rho \beta \Delta T$$

The negative sign means that if β is positive, any increase in temperature causes the density to decrease and vice versa. ◊

(b) We apply the equation $\beta = -\dfrac{\Delta \rho}{\rho \Delta T}$ for the specific case of water:

$$\beta = -\frac{(1.0000 \text{ g / cm}^3 - 0.9997 \text{ g / cm}^3)}{(1.0000 \text{ g / cm}^3)(4.00 \,°\text{C} - 10.0 \,°\text{C})}$$

Solving, we find that

$$\beta = 5.00 \times 10^{-5} \,°\text{C}^{-1} \qquad ◊$$

45. A vertical cylinder of cross-sectional area A is fitted with a tight-fitting, frictionless piston of mass m (Fig. P16.45). (a) If n moles of an ideal gas is in the cylinder at a temperature of T, what is the height h at which the piston is in equilibrium under its own weight? (b) What is the value for h if $n = 0.200$ mol, $T = 400$ K, $A = 0.00800$ m^2 and $m = 20.0$ kg?

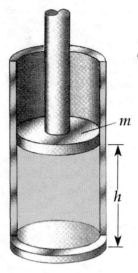

Solution

(a) We suppose that the air above the piston remains at atmospheric pressure P_0. Model the piston as a particle in equilibrium.

Figure P16.45

Then $\quad \Sigma F_y = ma_y \quad$ yields $\quad -P_0 A - mg + PA = 0$

where P is the pressure exerted by the gas contained.

Noting that $V = Ah$, and that n, T, m, g, A, and P_0 are given,

$PV = nRT$ becomes $\qquad\qquad P = \dfrac{nRT}{Ah}$

and $\qquad\qquad -P_0 A - mg + \dfrac{nRT}{Ah} A = 0$

$$h = \frac{nRT}{P_0 A + mg} \qquad\qquad \Diamond$$

(b) $\quad h = \dfrac{(0.200 \text{ mol})(8.315 \text{ J / mol} \cdot \text{K})(400 \text{ K})}{\left(1.013 \times 10^5 \text{ N / m}^2\right)\left(0.00800 \text{ m}^2\right) + (20.0 \text{ kg})\left(9.80 \text{ m / s}^2\right)}$

$h = \dfrac{665 \text{ N} \cdot \text{m}}{810 \text{ N} + 196 \text{ N}} = 0.661 \text{ m} \qquad\qquad \Diamond$

Chapter 17

Energy in Thermal Processes:
The First Law of Thermodynamics

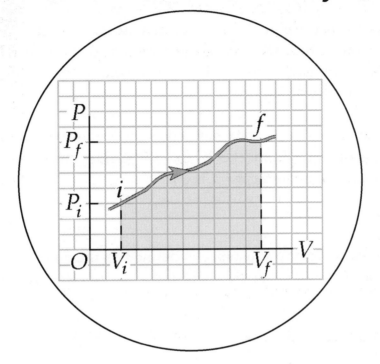

INTRODUCTION

This chapter focuses on the concept of heat, the first law of thermodynamics, thermal processes by which energy is transferred, and some important applications. The first law of thermodynamics is a reduced version of the continuity equation for energy, Equation 6.20. The first law places no restrictions on the types of energy conversions that can occur. Furthermore, it shows a commonality between heat and work. According to the first law, a system's internal energy can be increased either by energy transfer by heat to the system or by work done on the system. An important difference between heat and work is not evident from the first law; it is possible to transfer energy into a system by work and have it transfer out completely by heat, but it is impossible to transfer energy into a system undergoing a cyclic process by heat and have the energy transfer out completely by work. This difference is described by the second law of thermodynamics, in Chapter 18.

NOTES FROM SELECTED CHAPTER SECTIONS

17.1 Heat and Internal Energy

When two systems at different temperatures are in contact with each other, energy will transfer between them until they reach the same temperature (that is, until they are in thermal equilibrium with each other). The term heat refers to energy transfer as a consequence of a temperature difference.

One unit of heat is the calorie (cal), defined as the amount of heat necessary to increase the temperature of 1 g of water from 14.5 °C to 15.5 °C. The mechanical equivalent of heat, first measured by Joule, is given by 1 cal = 4.186 J.

17.2 Specific Heat

The **specific heat,** c, of any substance is defined as the amount of energy required to increase the temperature of 1 kg of that substance by one Celsius degree. Its units are $J/kg \cdot °C$.

17.3 Latent Heat and Phase Changes

The **heat of fusion** is the amount of energy required to melt (or freeze) 1 kg of a specific substance, with no temperature change in the substance. The **heat of vaporization** characterizes the liquid-to-gas change in a similar manner. Both of these parameters have units of J/kg.

17.4 Work in Thermodynamic Processes

The work done on a gas as its volume changes from V_i to V_f is given by Eq. 17.7,

$$W = -\int_i^f P\,dV$$

A **PV diagram** is a graphical representation which shows the path followed by a gas as it progresses from an initial to a final state. The work done on a gas in a process that takes the gas between initial and final states is the **negative of the area** under the curve on a PV diagram. Figure 17.1 illustrates the case for compression of a gas.

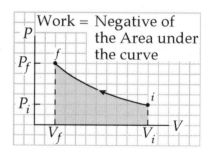

Figure 17.1

If a gas is compressed, $V_f < V_i$, and the work on the gas is positive. If the gas expands, $V_f > V_i$, the work on the gas is negative, and the gas does positive work on the piston. If the gas expands **at constant pressure**, called an isobaric process, then the work done on the gas is $W = -P(V_f - V_i)$.

The work done on a system depends on the process by which the system goes from the initial to the final state. In other words, the work done depends on the initial, final, and intermediate states of the system.

17.5 The First Law of Thermodynamics

In the first law of thermodynamics, $\Delta E_{int} = Q + W$, Q is the energy added to the system by heat and W is the work done on the system. Note that by convention, Q is **positive** when energy enters the system by heat and **negative** when energy is removed from the system by heat. Work, W, is positive when work is done on the gas (e.g. compression at constant temperature) and negative when work is done by the gas (e.g. expansion at constant pressure). The initial and final states must be **equilibrium** states; however, the intermediate states often are not equilibrium states since the thermodynamic coordinates may be impossible to determine during the thermodynamic process. For an **infinitesimal change of the system**, the **first law of thermodynamics** becomes $dE_{int} = dQ + dW$. We note that dQ and dW **are not exact differentials**, since both Q and W are not functions of the system coordinates. That is, both Q and W depend on the **path** taken between the initial and final equilibrium states, during which time the system interacts with its environment. On the other hand, dE_{int} is an **exact differential** and the internal energy E_{int} is a **state function**.

17.6 Some Applications of the First Law of Thermodynamics

An **isolated system** is one that does not interact with its surroundings. In such a system, $Q = W = 0$, so it follows from the first law that $\Delta E_{int} = 0$. That is, the internal energy of an isolated system cannot change.

A **cyclic process** is one that originates and ends up at the same state. In this situation, $\Delta E_{int} = 0$, so from the first law we see that $Q = -W$. That is, the work done per cycle equals the energy added by heat to the system per cycle. This is important to remember when dealing with heat engines in the next chapter.

An **adiabatic process** is a process in which no energy enters or leaves the system by heat; that is, $Q = 0$. The first law applied to this process gives $\Delta E_{int} = W$. A system may undergo an adiabatic process if it is thermally insulated from its surroundings, or if the process is so rapid that negligible energy has time to flow by heat.

An **isobaric process** is a process that occurs at constant pressure.

An **isovolumetric process** is one that occurs at constant volume. By definition, $W = 0$ for such a process (since $dV = 0$), so from the first law it follows that $\Delta E_{int} = Q$. That is, all of the energy added by heat to a system kept at constant volume goes into increasing the internal energy of the system.

A process that occurs at constant temperature is called an **isothermal process**, and a plot of P versus V at constant temperature for an ideal gas yields a hyperbolic curve called an **isotherm**. The internal energy of an ideal gas is a function of temperature only. Hence, in an isothermal process of an ideal gas, $\Delta E_{int} = 0$.

17.10 Energy Transfer Mechanisms in Thermal Processes

Three types of energy transfer are often related to temperature changes. These are (1) conduction, (2) convection, and (3) radiation.

Conduction is an energy transfer by heat. It occurs when there is a **temperature gradient** across the body. That is, conduction occurs only when the body's temperature is **not** uniform. For example, if you hold one end of a metal rod in a flame, energy will flow from the hot end to the colder end. If the flow is along x (that is, along the rod), and we define the **temperature gradient** as dT/dx, then in a time dt a quantity of energy dQ will flow by heat along the rod. The rate of energy flow along the rod is proportional to the cross-sectional area of the rod, the **temperature gradient**, and k, the thermal conductivity of the material of which the rod is made.

When energy transfer occurs as the result of the motion of material, such as the mixing of hot and cold fluids, the process is referred to as **convection**. Convection is used in conventional hot-air and hot-water heating systems. Convection currents produce changes in weather conditions when warm and cold air masses mix in the atmosphere.

Energy transfer by **radiation** is the result of the continuous emission of electromagnetic radiation by all bodies.

EQUATIONS AND CONCEPTS

Every substance requires a unique quantity of energy to change the temperature of 1 kg of the substance by 1 °C. This quantity is a measure of the **specific heat** of the substance.

$$c \equiv \frac{Q}{m \Delta T} \tag{17.2}$$

The energy Q that must be transferred to a system of mass m from its surroundings to produce a temperature change ΔT varies with the substance.

$$Q = mc\Delta T \tag{17.3}$$

A substance may undergo a phase change when energy is transferred between the substance and its surroundings. The energy Q required to change the phase of a mass m depends on a property of the material called the **latent heat**, L. The phase change process occurs at constant temperature. The value of L depends on the nature of the phase change and the thermal properties of the substance. Choose the positive sign when energy is transferred into the system to change solid to liquid or liquid to gas. Choose the negative sign when energy leaves the system during condensation or freezing.

$$Q = \pm mL \tag{17.5}$$

The **work done on a gas** that undergoes an expansion or compression from volume V_i to volume V_f depends on the path between the initial and final states. The pressure is generally not constant, so you must exercise care in evaluating W using this equation. In general, the work done is the negative of the area under the PV curve bounded by V_i and V_f, and the pressure function P.

$$W = -\int_{V_i}^{V_f} P\,dV \qquad (17.7)$$

The **first law of thermodynamics** is a version of the law of conservation of energy that describes changes in internal energy. It states that the **change** in internal energy of a system, ΔE_{int}, equals the quantity $Q + W$, where Q is the energy transferred by heat to the system and W is the work done on it. The quantity $Q + W$ is independent of the path taken between the initial and final states.

$$\Delta E_{int} = Q + W \qquad (17.8)$$

In an **adiabatic process**, $Q = 0$, and the change in internal energy equals the work done **on** the gas.

$$\Delta E_{int} = W \qquad (17.10)$$

In an isovolumetric (constant volume) process, zero work is done and all energy transferred to the system by heat increases the internal energy.

$$\Delta E_{int} = Q \qquad (17.11)$$

An **isothermal process** is one that occurs at constant temperature. In an isothermal process for an ideal gas, the internal energy is constant. Any energy transferred to the gas by work is transferred by heat from the gas to its surroundings.

$$W = -Q = -nRT \ln\left(\frac{V_f}{V_i}\right) \qquad (17.12)$$

The **internal energy** E_{int} of N molecules (or n moles) of a monatomic ideal gas is proportional to the absolute temperature.

$$E_{int} = \frac{3}{2} N k_B T = \frac{3}{2} nRT \qquad (17.15)$$

or

$$E_{int} = n C_V T \qquad (17.18)$$

If we apply the first law of thermodynamics to a monatomic **ideal gas** in which energy is transferred by heat at **constant volume**, we find that the **molar specific heat at constant volume** is equal to $\frac{3}{2} R$.

$$C_V = \frac{3}{2} R \quad \text{(monatomic only)} \qquad (17.17)$$

If energy is transferred by heat to an ideal gas at **constant pressure**, the first law of thermodynamics shows that the **molar specific heat at constant pressure** is greater than the molar specific heat at constant volume by an amount R.

$$C_P - C_V = R \qquad \text{(any ideal gas)} \qquad (17.21)$$

or

$$C_P = \frac{5}{2} R \qquad \text{(monatomic)}$$

The **ratio of the specific heat** γ is a dimensionless quantity that is equal to 1.67 for a **monatomic ideal gas**.

$$\gamma = \frac{C_P}{C_V} \qquad \text{(always)} \qquad (17.22)$$

$$\gamma = \frac{C_P}{C_V} = 1.67 \qquad \text{(monatomic)}$$

If an **ideal gas** undergoes an **adiabatic expansion**, and we assume $PV = nRT$ is valid at any time, then the pressure and volume of the gas obey Equation 17.24.

$$PV^\gamma = \text{constant} \qquad (17.24)$$

SUGGESTIONS, SKILLS, AND STRATEGIES

Many applications of the first law of thermodynamics deal with the work done on a system that undergoes a change in state. As a gas is taken from a state with initial pressure P_i and volume V_i, to a state with final pressure P_f and volume V_f, the work can be calculated if the process can be drawn on a PV diagram as in Figure 17.2. The work input during the expansion is given by the integral expression

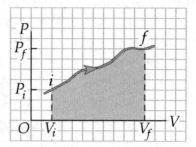

Figure 17.2

$$W = -\int_{V_i}^{V_f} P\,dV$$

which numerically represents the **negative of the area** under the PV curve (the shaded region) shown in Figure 17.2. It is important to recognize that the work **depends on the path taken as the gas goes from *i* to *f*.** That is, W depends on the specific manner in which the pressure P changes during the process.

Problem-Solving Strategy: Calorimetry Problems

If you are having difficulty with calorimetry problems, consider the following factors:

- Be sure your units are consistent throughout. For instance, if you are using specific heats in cal / g · °C, be sure that masses are in grams and temperatures are Celsius throughout.

- Energy losses and gains are found by using $Q = mc\Delta T$ only for those intervals in which no phase changes are occurring. The equations $Q = \pm mL_f$ and $Q = \pm mL_v$ are to be used only when phase changes **are** taking place.

- Often sign errors occur in calorimetry equations. Remember the negative sign in the equation $Q_{\text{cold}} = -Q_{\text{hot}}$.

Calculating Specific Heats

It is important remember that for gases, two different values of molar heat capacity can be calculated: C_P, molar heat capacity at constant pressure and C_V, molar heat capacity at constant volume. For any ideal gas $C_P - C_V = R$, the universal gas constant and the ratio of $(C_P / C_V) = \gamma$ has a value which depends on the nature of the gas (see Table 17.3, p. 602 of the text). The two equations above can be combined to find the following expressions for C_V and C_P:

$$C_V = \frac{R}{\gamma - 1} \qquad \text{and} \qquad C_P = \frac{\gamma R}{\gamma - 1}$$

These may be used for ideal gases that are not monatomic.

REVIEW CHECKLIST

▷ Understand the concepts of heat, internal energy, and thermodynamic processes.

▷ Define and discuss the calorie, specific heat, and latent heat.

▷ Understand how work is defined when a system undergoes a change in state, and the fact that work (like heat) depends on the path taken by the system. You should also know how to sketch processes on a PV diagram, and calculate work using these diagrams.

▷ State the first law of thermodynamics ($\Delta E_{int} = Q + W$), and explain the meaning of the three energy terms related by this statement.

▷ Discuss the implications of the first law of thermodynamics as applied to (a) an isolated system, (b) a cyclic process, (c) an adiabatic process, and (d) an isothermal process.

▷ Recognize that the internal energy of an ideal gas is proportional to the absolute temperature, and be able to derive the specific heat of an ideal gas at constant volume from the first law of thermodynamics.

▷ Define an adiabatic process, and be able to derive the expression $PV^\gamma = constant$, which applies to an adiabatic process for an ideal gas.

ANSWERS TO SELECTED CONCEPTUAL QUESTIONS

3. Using the first law of thermodynamics, explain why the **total** energy of an isolated system is always constant.

Answer

The first law of thermodynamics says that the net change in internal energy of a system is equal to the energy added by heat, plus the work done on the system.

$$\Delta E_{int} = Q + W$$

However, an isolated system is defined as an object or set of objects for which there is no exchange of energy with its surroundings. In the case of the first law of thermodynamics, this means that $Q = W = 0$, so the change in internal energy of the system at all times must be zero.

As it stays constant in amount, an isolated system's energy may change from one form to another or move from one object to another within the system. For example, a "bomb calorimeter" is a closed system that consists of a sturdy steel container, a water bath, an item of food, and oxygen. The food is burned in the presence of an excess of oxygen, and chemical energy is converted to internal energy. In the process of oxidation, energy is transferred by heat to the water bath, raising its temperature. The change in temperature of the water bath is used to determine the caloric content of the food. The process works specifically **because** the total energy remains unchanged in a closed system.

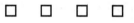

7. What is wrong with the statement: "given any two bodies, the one with the higher temperature contains more heat"?

Answer

The statement shows a misunderstanding of the concept of heat. Heat is a process by which energy is transferred, not a form of energy that is held or contained. If you wish to speak of energy that is "contained", you speak of **internal energy**, not **heat**.

Further, even if the statement used the term "internal energy", it would still be incorrect, since the effects of specific heat and mass are both ignored. A 1-kg mass of water at 20 °C has more internal energy than a 1-kg mass of air at 30 °C. Similarly, the earth has far more internal energy than a drop of molten titanium metal.

Correct statements would be: (1) "given any two bodies in thermal contact, the one with the higher temperature will transfer energy to the other by heat". (2) "given any two bodies of equal mass, the one with the higher product of absolute temperature and specific heat contains more internal energy".

□ □ □ □

10. The air temperature above coastal areas is profoundly influenced by the large specific heat of water. One reason is that the energy released when 1 m^3 of water cools by 1.0 °C will raise the temperature of an enormously larger volume of air by 1.0 °C. Estimate this volume of air. The specific heat of air is approximately 1.0 kJ/kg · °C. Take the density of air to be 1.3 kg/m^3.

Answer:

The mass of one cubic meter of water is specified by its density,

$$m = \rho V = \left(1.00 \times 10^3 \ \text{kg/m}^3\right)\left(1 \ \text{m}^3\right) = 1 \times 10^3 \ \text{kg}$$

When one cubic meter of water cools by 1 °C it releases energy

$$Q = mc\Delta T = \left(1 \times 10^3 \ \text{kg}\right)\left(4186 \ \text{J/kg} \cdot °\text{C}\right)\left(-1 \ °\text{C}\right) = -4 \times 10^6 \ \text{J}$$

where the negative sign represents heat output. When $+4 \times 10^6$ J is transferred to the air, raising its temperature 1 °C, the volume of the air is given by $Q = mc\Delta T = \rho V c \Delta T$:

$$V = \frac{Q}{\rho c \Delta T} = \frac{4 \times 10^6 \ \text{J}}{\left(1.3 \ \text{kg/m}^3\right)\left(1 \times 10^3 \ \text{J/kg} \cdot °\text{C}\right)\left(1 \ °\text{C}\right)} = 3 \times 10^3 \ \text{m}^3$$

The volume of the air is thousands of times larger than the volume of the water.

□ □ □ □

SOLUTIONS TO SELECTED END-OF-CHAPTER PROBLEMS

5. A 1.50-kg iron horseshoe initially at 600 °C is dropped into a bucket containing 20.0 kg of water at 25.0 °C. What is the final temperature? (Ignore the specific heat of the container, and assume that a negligible amount of water boils away.)

Solution

Even though the horseshoe is much hotter than the water, the mass of the water is significantly greater, so we might expect the water temperature to rise less than 10 °C. Energy transfer by heat proceeds until the horseshoe and the water attain the same final temperature T. The energy lost by the iron will be gained by the water, and from this energy exchange the final temperature of the water can be calculated.

$$Q_{iron} = -Q_{water} \qquad \text{or} \qquad (mc\,\Delta T)_{iron} = -(mc\,\Delta T)_{water}$$

$$m_{Fe}c_{Fe}(T - 600\ °C) = -m_w c_w (T - 25.0\ °C)$$

$$T = \frac{m_w c_w (25.0\ °C) + m_{Fe} c_{Fe}(600\ °C)}{m_w c_w + m_{Fe} c_{Fe}}$$

$$T = \frac{(20.0\ \text{kg})(4186\ \text{J/kg} \cdot °C)(25.0\ °C) + (1.50\ \text{kg})(448\ \text{J/kg} \cdot °C)(600\ °C)}{(20.0\ \text{kg})(4186\ \text{J/kg} \cdot °C) + (1.50\ \text{kg})(448\ \text{J/kg} \cdot °C)}$$

$$T = 29.6\ °C \qquad\qquad\qquad \Diamond$$

The temperature only rose about 5 °C, so our answer seems reasonable. The specific heat of the water is about 10 times greater than that of the iron, so this effect also reduces the change in water temperature. In this problem we assumed that a negligible amount of water boiled away, but in reality the final temperature of the water would be somewhat less than we calculated, since some of the energy transferred to the water by heat would vaporize a bit of water.

11. A 3.00-g lead bullet at 30.0 °C is fired at a speed of 240 m/s into a large block of ice at 0 °C, in which it becomes embedded. What quantity of ice melts?

Solution The amount of ice that melts is probably small, maybe only a few grams based on the size, speed, and initial temperature of the bullet.

We use the energy version of the isolated system model for the bullet-plus-ice-block system. We assume that all of the initial kinetic and excess internal energy of the bullet goes into internal energy to melt the ice, the mass of which can be found from the latent heat of fusion. To reach thermal equilibrium the bullet must cool to 0 °C. The energy lost by the bullet equals the energy gained by the ice: $\left|\Delta K_b\right| + \left|Q_b\right| = Q_{ice}$

$$\frac{1}{2}m_b v_b^2 + m_b c_{Pb}\left|\Delta T\right| = m_{ice}L_f \qquad \text{or} \qquad m_{ice} = m_b\left(\frac{\frac{1}{2}v_b^2 + c_{Pb}\left|\Delta T\right|}{L_f}\right)$$

$$m_{ice} = \left(3.00\times10^{-3}\text{ kg}\right)\left(\frac{\frac{1}{2}(240\text{ m/s})^2 + (128\text{ J/kg}\cdot°\text{C})(30.0\text{ °C})}{3.33\times10^5\text{ J/kg}}\right)$$

$$m_{ice} = \frac{86.4\text{ J} + 11.5\text{ J}}{3.33\times10^5\text{ J/kg}} = 2.94\times10^{-4}\text{ kg} = 0.294\text{ g} \qquad \lozenge$$

The amount of ice that melted is less than gram, which agrees with our prediction. It appears that most of the energy used to melt the ice comes from kinetic energy of the bullet (88%), while the excess internal energy of the bullet only contributes 12% to melt the ice. Small chips of ice probably fly off when the bullet makes impact. Some of the energy is transferred to their kinetic energy, so in reality, the amount of ice that would melt should be less than what we calculated. If the block of ice were colder than 0 °C (as is usually in the case), then the melted ice would refreeze.

---

13. In an insulated vessel, 250 g of ice at 0 °C is added to 600 g of water at 18.0 °C. (a) What is the final temperature of the system? (b) How much ice remains when the system reaches equilibrium?

Solution

(a) When 250 g of ice is melted,

$$Q_f = mL_f = (0.250 \text{ kg})(3.33 \times 10^5 \text{ J/kg}) = 83.3 \text{ kJ}$$

The energy released when 600 g of water cools from 18.0 °C to 0 °C is

$$|Q| = |mc\Delta T| = (0.600 \text{ kg})(4186 \text{ J/kg} \cdot °\text{C})(18.0 °\text{C}) = 45.2 \text{ kJ}$$

Since the energy required to melt 250 g of ice at 0 °C **exceeds** the energy released by cooling 600 g of water from 18.0 °C to 0 °C, the final temperature of the system (water + ice) must be 0 °C. ◊

(b) The energy released by the water (45.2 kJ) will melt a mass of ice m, where $Q = mL_f$.

Solving for the mass, $\qquad\qquad m = \dfrac{Q}{L_f} = \dfrac{45.2 \times 10^3 \text{ J}}{3.33 \times 10^5 \text{ J/kg}} = 0.136 \text{ kg}$

Therefore, the ice remaining is $\qquad\qquad m' = 0.250 \text{ kg} - 0.136 \text{ kg} = 0.114 \text{ kg}$ ◊

---

17. A sample of ideal gas is expanded to twice its original volume of 1.00 m^3 in a quasi-static process for which $P = \alpha V^2$, with $\alpha = 5.00$ atm/m^6, as shown in Fig. P17.17. How much work was done on the expanding gas?

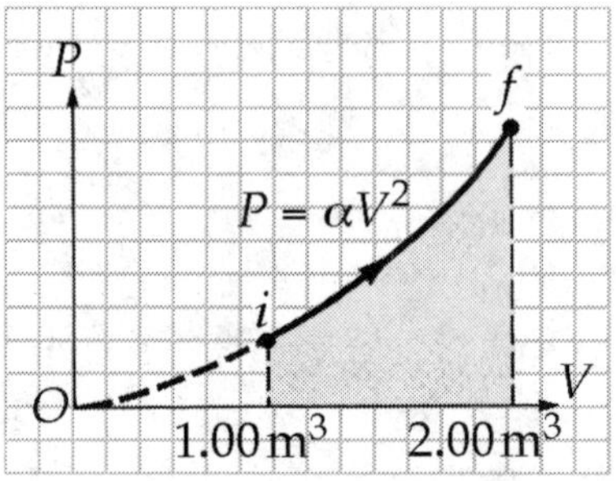

Solution $\qquad W = -\displaystyle\int_{V_i}^{V_f} P\,dV$

$$V_f = 2V_i = 2(1.00 \text{ m}^3) = 2.00 \text{ m}^3$$

Figure P17.17

The work done on the gas is the negative of the area under the curve $P = \alpha V^2$, from V_i to V_f.

$$W = -\int_{V_i}^{V_f} \alpha V^2\,dV = -\tfrac{1}{3}\alpha\left(V_f^3 - V_i^3\right)$$

$$W = -\tfrac{1}{3}(5.00 \text{ atm/m}^6)(1.013 \times 10^5 \text{ Pa/atm})\left[(2.00 \text{ m}^3)^3 - (1.00 \text{ m}^3)^3\right] = -1.18 \times 10^6 \text{ J}$$ ◊

---

21. A thermodynamic system undergoes a process in which its internal energy decreases by 500 J. At the same time, 220 J of work is done on the system. Find the energy transferred to or from it by heat.

Solution $\Delta E_{int} = Q + W$, where W is positive, because work is done **on** the system:

$$Q = \Delta E_{int} - W = -500 \text{ J} - 220 \text{ J} = -720 \text{ J}$$

+720 J of energy is transferred **from** the system by heat.   ◊

---

25. An ideal gas initially at 300 K undergoes an isobaric expansion at 2.50 kPa. If the volume increases from 1.00 m^3 to 3.00 m^3 and 12.5 kJ is transferred to the gas by heat, what are (a) the change in its internal energy and (b) its final temperature?

Solution   We use the energy version of the nonisolated system model.

(a) $$\Delta E_{int} = Q + W \qquad \text{where} \qquad W = -P\Delta V$$

so that $$\Delta E_{int} = Q - P\Delta V$$

$$\Delta E_{int} = 1.25 \times 10^4 \text{ J} - \left(2.50 \times 10^3 \text{ N/m}^2\right)\left(3.00 \text{ m}^3 - 1.00 \text{ m}^3\right) = 7500 \text{ J} \quad ◊$$

(b)   Since $\dfrac{V_1}{T_1} = \dfrac{V_2}{T_2}$, $T_2 = \left(\dfrac{V_2}{V_1}\right)T_1 = \left(\dfrac{3.00 \text{ m}^3}{1.00 \text{ m}^3}\right)(300 \text{ K}) = 900 \text{ K}$ ◊

---

27. A 2.00-mol sample of helium gas initially at 300 K and 0.400 atm is compressed isothermally to 1.20 atm. Assuming the helium behaves as an ideal gas, find (a) the final volume of the gas, (b) the work done on the gas, and (c) the energy transferred by heat.

Solution

(a) Rearranging $PV = nRT$

we get $V_i = \dfrac{nRT}{P_i} = \dfrac{(2.00 \text{ mol})(8.315 \text{ J/mol} \cdot \text{K})(300 \text{ K})}{(0.400 \text{ atm})(1.013 \times 10^5 \text{ Pa/atm})}\left(\dfrac{1 \text{ Pa}}{\text{N/m}^2}\right) = 0.123 \text{ m}^3$

For isothermal compression, PV is constant, so $P_i V_i = P_f V_f$:

$$V_f = V_i\left(\dfrac{P_i}{P_f}\right) = (0.123 \text{ m}^3)\left(\dfrac{0.400 \text{ atm}}{1.20 \text{ atm}}\right) = 0.0410 \text{ m}^3 \qquad \lozenge$$

(b) $W = -\int P\,dV :$ $\qquad W = -\int \dfrac{nRT}{V}\,dV = -nRT\ln\left(\dfrac{V_f}{V_i}\right) = -(4989 \text{ J})\ln\left(\tfrac{1}{3}\right) = +5.48 \text{ kJ} \qquad \lozenge$

(c) $\Delta E_{\text{int}} = 0 = Q + W :$ $\quad Q = -5.48 \text{ kJ}$ $\qquad\qquad\qquad\qquad\qquad\qquad\qquad\quad \lozenge$

---

31. One mole of hydrogen gas is heated at constant pressure from 300 K to 420 K. Calculate (a) the energy transferred to the gas by heat, (b) the increase in its internal energy, and (c) the work done on the gas.

Solution Since this is a constant-pressure process, $Q = nC_P\Delta T$

(a) The temperature rises by $\qquad\qquad \Delta T = 420 \text{ K} - 300 \text{ K} = 120 \text{ K}$

$$Q = (1.00 \text{ mol})(28.8 \text{ J/mol} \cdot \text{K})(120 \text{ K}) = 3.46 \text{ kJ} \qquad \lozenge$$

(b) For any system, $\qquad\qquad\qquad \Delta E_{\text{int}} = nC_V\Delta T$

so $\qquad\qquad\qquad\qquad\qquad \Delta E_{\text{int}} = (1.00 \text{ mol})(20.4 \text{ J/mol} \cdot \text{K})(120 \text{ K}) = 2.45 \text{ kJ} \quad \lozenge$

(c) $\Delta E_{\text{int}} = Q + W \qquad$ so $\qquad W = \Delta E_{\text{int}} - Q = 2.45 \text{ kJ} - 3.46 \text{ kJ} = -1.01 \text{ kJ} \qquad \lozenge$

---

37. A 2.00-mol sample of an ideal gas with $\gamma = 1.40$ expands slowly and adiabatically from a pressure of 5.00 atm and a volume of 12.0 L to a final volume of 30.0 L. (a) What is the final pressure of the gas? (b) What are the initial and final temperatures? (c) Find Q, W and ΔE_{int}.

Solution

(a) $\quad P_i V_i^{\gamma} = P_f V_f^{\gamma}$

$$P_f = P_i \left(\frac{V_i}{V_f} \right)^{\gamma} = (5.00 \text{ atm}) \left(\frac{12.0 \text{ L}}{34.0 \text{ L}} \right)^{1.40} = 1.39 \text{ atm} \qquad \lozenge$$

(b) $\quad T_i = \dfrac{P_i V_i}{nR} = \dfrac{(5.00 \text{ atm})(1.013 \times 10^5 \text{ Pa/atm})(12.0 \times 10^{-3} \text{ m}^3)}{(2.00 \text{ mol})(8.315 \text{ N} \cdot \text{m/mol} \cdot \text{K})} = 365 \text{ K}$

$$T_f = \frac{P_f V_f}{nR} = 253 \text{ K} \qquad \lozenge$$

(c) This is an adiabatic process, so by the definition $\qquad\qquad Q = 0 \qquad \lozenge$

For any process, $\qquad \Delta E_{int} = n C_V \Delta T$

and for this gas, $\qquad C_V = \dfrac{R}{\gamma - 1} = \dfrac{5}{2} R$

Thus, $\qquad\qquad \Delta E_{int} = \dfrac{5}{2} (2.00 \text{ mol})(8.315 \text{ J/mol} \cdot \text{K})(253 \text{ K} - 365 \text{ K}) = -4660 \text{ J} \qquad \lozenge$

Now, $\qquad\qquad W = \Delta E_{int} - Q = -4660 \text{ J} - 0 = -4660 \text{ J} \qquad \lozenge$

Note that in this case the work done on the gas is negative, so positive work is done by the gas.

43. The **heat capacity** of a sample of a substance is the product of the mass of the sample and the specific heat of the substance. Consider 2.00 mol of an ideal diatomic gas. Find the total heat capacity at constant volume and at constant pressure (a) if the molecules rotate but do not vibrate and (b) if the molecules both rotate and vibrate.

Solution   We use the kinetic theory structural model of an ideal gas.

(a)   Count degrees of freedom. A diatomic molecule oriented along the y axis can possess energy by moving in x, y and z directions and by rotating around x and z axes. Rotation around the y axis does not represent an energy contribution because the moment of inertia of the molecule about this axis is essentially zero. The molecule will have an average energy $\frac{1}{2}k_BT$ for each of these five degrees of freedom.

The gas will have internal energy $E_{int} = N\left(\frac{5}{2}\right)k_BT = nN_A\left(\frac{5}{2}\right)\left(\frac{R}{N_A}\right)T = \frac{5}{2}nRT$

so the constant-volume heat capacity of the whole sample is

$$\frac{\Delta E_{int}}{\Delta T} = \frac{5}{2}nR = \frac{5}{2}(2.00 \text{ mol})(8.315 \text{ J/mol}\cdot\text{K}) = 41.6 \text{ J/K} \qquad \Diamond$$

For one mole, $C_P = C_V + R$;   For the sample, $nC_P = nC_V + nR$

With P constant, $nC_P = 41.6 \text{ J/K} + (2.00 \text{ mol})(8.315 \text{ J/mol}\cdot\text{K}) = 58.2 \text{ J/K} \qquad \Diamond$

(b)   Vibration adds a degree of freedom for kinetic energy and a degree of freedom for elastic energy. Now the molecule's average energy is $\frac{7}{2}k_BT$,

the sample's internal energy is $N\left(\frac{7}{2}\right)k_BT = \frac{7}{2}nRT$

and the sample's constant-volume heat capacity is

$$\frac{7}{2}nR = \frac{7}{2}(2.00 \text{ mol})(8.315 \text{ J/mol}\cdot\text{K}) = 58.2 \text{ J/K} \qquad \Diamond$$

At constant pressure, its heat capacity is

$$nC_V + nR = 58.2 \text{ J/K} + (2.00 \text{ mol})(8.315 \text{ J/mol}\cdot\text{K}) = 74.8 \text{ J/K} \qquad \Diamond$$

47. A bar of gold is in thermal contact with a bar of silver of the same length and area (Fig. P17.47). One end of the compound bar is maintained at 80.0 °C, and the opposite end is at 30.0 °C. When the energy transfer reaches steady state, what is the temperature at the junction?

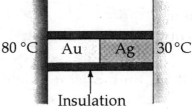

80 °C   Au   Ag   30°C

Insulation

Figure P17.47

Solution

Call the gold bar Object 1 and the silver bar Object 2. Each is a nonisolated system in steady state. When energy transfer by heat reaches a steady state, the flow rate through each will be the same:

$$\mathcal{P}_1 = \mathcal{P}_2 \qquad \text{or} \qquad \frac{k_1 A_1 \Delta T}{L_1} = \frac{k_2 A_2 \Delta T}{L_2}$$

In this case, $\qquad L_1 = L_2 \qquad$ and $\qquad A_1 = A_2$

so $\qquad k_1 \Delta T_1 = k_2 \Delta T_2$

Let T_3 be the temperature at the junction; then $\qquad k_1 \left(80.0\ °C - T_3\right) = k_2\left(T_3 - 30.0\ °C\right)$

Rearranging, we find $\qquad T_3 = \dfrac{\left(80.0\ °C\right)k_1 + \left(30.0\ °C\right)k_2}{k_1 + k_2}$

$$T_3 = \frac{\left(80.0\ °C\right)\left(314\ W/m \cdot °C\right) + \left(30.0\ °C\right)\left(427\ W/m \cdot °C\right)}{\left(314\ W/m \cdot °C\right) + \left(427\ W/m \cdot °C\right)}$$

$$T_3 = 51.2\ °C \qquad\qquad\qquad ◊$$

61. A solar cooker consists of a curved reflecting surface that concentrates sunlight onto the object to be warmed (Fig. P17.61). The solar power per unit area reaching the Earth's surface at the location is 600 W/m^2, and the cooker has a diameter of 0.600 m. Assume that 40.0% of the incident energy is transferred to 0.500 L of water in an open container, initially at 20.0 °C. How long does it take to completely boil away the water? (Ignore the specific heat of the container.)

Figure P17.61

Solution   If we point the axis of the reflecting surface toward the Sun, the power incident on the solar collector is

$$\mathcal{P}_i = IA = \left(600 \ \text{W/m}^2\right)\left[\pi(0.300 \ \text{m})^2\right] = 170 \ \text{W}$$

For a 40.0% reflector, the collected power is

$$\mathcal{P}_i = (0.400)(170 \ \text{W}) = 67.9 \ \text{W} = 67.9 \ \text{J/s}$$

The total energy required to increase the temperature of the water to the boiling point and to evaporate it is

$$Q = mc\,\Delta T + mL_v$$

$$Q = \left(0.500 \ \text{kg}\right)\left(4186 \ \text{J/kg} \cdot {}^\circ\text{C}\right)(80.0 \ {}^\circ\text{C}) + \left(0.500 \ \text{kg}\right)\left(2.26 \times 10^6 \ \text{J/kg}\right)$$

$$Q = 1.30 \times 10^6 \ \text{J}$$

The time required is

$$\Delta t = \frac{Q}{\mathcal{P}_c} = \frac{1.30 \times 10^6 \ \text{J}}{67.9 \ \text{J/s}} = 1.91 \times 10^4 \ \text{s} = 5.31 \ \text{h} \qquad \lozenge$$

Chapter 18

Heat Engines, Entropy, and the Second Law of Thermodynamics

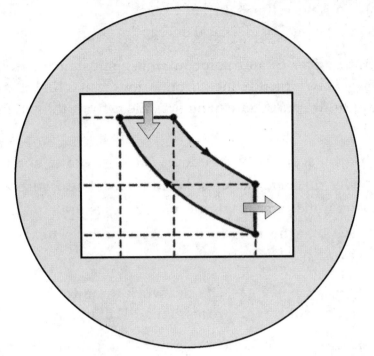

INTRODUCTION

The first law of thermodynamics, studied in Chapter 17, is a special case of the continuity equation for energy that is useful in studying thermal processes. The first law implies that energy cannot be created or destroyed, but it places no restrictions on the types of energy conversions that can occur. The second law of thermodynamics establishes which processes in nature do and which do not occur.

From an engineering viewpoint, perhaps the most important application of the second law of thermodynamics is the limited efficiency of heat engines. The second law says that no machine, operating in a cycle, can take energy into a system by heat and have the energy transfer out completely by work.

NOTES FROM SELECTED CHAPTER SECTIONS

18.1   Heat Engines and the Second Law of Thermodynamics

A heat engine is a device that takes in energy as heat and puts out energy in other useful forms, such as mechanical or electrical energy.

A practical heat engine carries some working substance through a cyclic process during which (1) energy is absorbed by heat from a source at a high temperature, (2) work is done by the engine, and (3) energy is expelled by heat from the engine to a reservoir at a lower temperature.

The engine absorbs a quantity of energy, $|Q_h|$, from a hot reservoir, does work $|W| = W_{eng}$, and then gives up energy $|Q_c|$ to a cold reservoir. Because the working substance goes through a cycle, its initial and final internal energies are equal, so $\Delta E_{int} = 0$. Hence, from the first law we see that **the net work, W_{eng}, done by a heat engine equals the net energy flowing into it by heat**.

If the working substance is a gas, **the net work done for a cyclic process is the area enclosed by the curve representing the process on a PV diagram.**

The **thermal efficiency**, e, of a heat engine is the ratio of the net work done by the engine to the energy absorbed by heat at the higher temperature during one cycle.

The **second law of thermodynamics** can be stated in several ways:

- **Clausius Statement:** Energy will not flow spontaneously from a cold object to a hot object. No thermodynamic process can occur whose only result is to transfer energy from a colder to a hotter body by heat. Such a (refrigeration) process is only possible if work is done on the system.
- **Kelvin-Planck Statement:** It is impossible for a cyclic thermodynamic process to occur whose only result is the complete conversion of energy extracted by heat from a hot reservoir into work. That is, **it is impossible to construct a heat engine that, operating in a cycle, produces no other effect than the absorption of energy by heat from a reservoir and the performance of an equal amount of work.** A heat engine must also release energy by heat into a cold reservoir. The efficiency of a heat engine must be less than 100%. The engine's energy output by heat is often called heat exhaust, wasted heat, rejected heat, expelled heat, or thermal pollution.
- The **entropy statement** is given in Section 18.8.

18.2 Reversible and Irreversible Processes

A process is **reversible** if the system passes from the initial to the final state through a succession of equilibrium states. Then, the process can be made to run in the opposite direction at any point by an infinitesimal change in conditions. Otherwise the process is **irreversible**.

18.3 The Carnot Engine

An ideal reversible cyclic process, called the **Carnot cycle,** is described in the PV diagram of Figure 18.1. The Carnot cycle consists of two adiabatic and two isothermal processes, all being reversible.

- The process $A \rightarrow B$ is an isotherm (constant T), during which time the gas expands at constant temperature T_h, and absorbs energy $|Q_h|$ by heat from the hot reservoir.

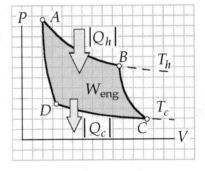

Figure 18.1

- The process $B \rightarrow C$ is an adiabatic expansion ($Q = 0$), during which time the gas expands and cools to a temperature T_c.

- The process $C \rightarrow D$ is a second isotherm, during which time the gas is compressed at constant temperature T_c, and gives up energy $|Q_c|$ by heat to the cold reservoir.

- The final process $D \rightarrow A$ is an adiabatic compression in which the gas temperature increases to a final temperature of T_h.

No working engine is 100% efficient, even when losses such as friction are neglected. One can determine the theoretical limit on the efficiency of a real engine by comparison with the ideal Carnot engine. A **reversible engine** is one that will operate with the same efficiency in the forward and reverse directions. The Carnot engine is one example of a reversible engine.

Carnot's theorems, which can be proved from the first and second laws of thermodynamics, can be stated as follows:

- **Theorem I.** No real (irreversible) engine can have an efficiency greater than that of a reversible engine operating between the same two temperatures.

- **Theorem II.** All reversible engines operating between T_h and T_c have the **same** efficiency, given by Equation 18.4.

A schematic diagram of a heat engine is shown in Figure 18.2a, where $|Q_h|$ is the energy extracted from the hot reservoir by heat at temperature T_h, $|Q_c|$ is the energy rejected to the cold reservoir by heat at temperature T_c, and W_{eng} is the work done by the engine.

18.4 Heat Pumps and Refrigerators

A refrigerator is a heat engine operating in reverse, as shown in Figure 18.2b. During one cycle of operation, the refrigerator absorbs energy $|Q_c|$ by heat from the cold reservoir, expels energy $|Q_h|$ by heat to the hot reservoir, and the work done on the system is $W = |Q_h| - |Q_c|$.

Hot Reservoir at T_h 　　　　　　　　　　　Hot Reservoir at T_h

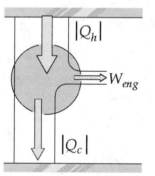
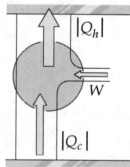

Cold Reservoir at T_c 　　　　　　　　　　Cold Reservoir at T_c

(a)　　Heat engine 　　　　　　　　(b)　　Refrigerator (heat pump)

Figure 18.2

18.6 – 18.8   Entropy

Entropy is a quantity used to measure the degree of **disorder** in a system. For example, the molecules of a gas in a container at a high temperature are in a more disordered state (higher entropy) than the same molecules at a lower temperature.

When energy is added by heat to a system in an incremental reversible process, dQ_r is **positive** and the entropy **increases**. When energy is removed, dQ_r is **negative** and the entropy **decreases**. Note that only **changes** in entropy are defined by Equation 18.8; therefore, the concept of entropy is most useful when a system undergoes a **change in its state.**

When using Equation 18.12 to calculate entropy changes, note that ΔS may be obtained even if the process is irreversible, since ΔS depends only on the initial and final equilibrium states, not on the path. In order to calculate ΔS for an irreversible process, you must devise a reversible process (or sequence of reversible processes) between the initial and final states, and compute dQ_r / T for the reversible process. The entropy change for the irreversible process is the **same** as that of the reversible process between the same initial and final equilibrium states.

The second law of thermodynamics can be stated in terms of entropy as follows: **The total entropy of an isolated system always increases in time if the system undergoes an irreversible process.** If an isolated system undergoes a **reversible** process, the total entropy **remains constant.**

EQUATIONS AND CONCEPTS

The net **work done** by a heat engine during one cycle equals the net energy flowing into the engine as heat.

$$W_{eng} = |Q_h| - |Q_c| \tag{18.1}$$

The **thermal efficiency**, e, of a heat engine is defined as the ratio of the net work done to the energy absorbed as heat during one cycle of the process.

$$e = \frac{W_{eng}}{|Q_h|} = 1 - \frac{|Q_c|}{|Q_h|} \tag{18.2}$$

No real engine operating between the temperatures T_c and T_h can be more efficient than an engine operating reversibly in a Carnot cycle between the same two temperatures; this basic limit of efficiency is called the **Carnot efficiency**.

$$e_c = 1 - \frac{T_c}{T_h} \qquad (18.4)$$

The effectiveness of a heat pump or refrigerator is described in terms of the **coefficient of performance**, COP. The COP of a heat pump, operating in the heating mode, is the ratio of the energy transferred as heat at the high temperature to the work done on the pump.

$$\text{COP (heating mode)} = \frac{|Q_h|}{W} \qquad (18.5)$$

The **coefficient of performance** of a refrigerator (or a heat pump operating in the cooling mode) is defined by the ratio of the energy absorbed by heat Q_c, to the work done. A good refrigerator has a high coefficient of performance.

$$\text{COP (cooling mode)} = \frac{|Q_c|}{W} \qquad (18.6)$$

If a system changes from one equilibrium state to another, under a reversible, quasi-static process, and a quantity of energy dQ_r is added (or removed) by heat at the absolute temperature T, the **change in entropy** is defined by the ratio dQ_r / T.

$$dS = \frac{dQ_r}{T} \qquad (18.8)$$

All systems tend toward disorder, and the entropy of the system is related to W, the number of possible microstates of the system.

$$S \equiv k_B \ln W \qquad (18.11)$$

The **change in entropy** of a system that undergoes a reversible process between the states i and f depends only on the properties of the initial and final equilibrium states.

$$\Delta S = \int_i^f \frac{dQ_r}{T} \text{ (reversible path)} \qquad (18.12)$$

The change in entropy of a system for any arbitrary **reversible cycle** is identically **zero**.

$$\oint \frac{dQ_r}{T} = 0 \qquad (18.13)$$

During a **free expansion**, the entropy of a gas **increases**.

$$\Delta S = nR\ln\left(\frac{V_f}{V_i}\right) \qquad (18.14)$$

REVIEW CHECKLIST

▷ Understand the basic principle of the operation of a heat engine, and be able to define and discuss the **thermal efficiency** of a heat engine.

▷ State the second law of thermodynamics, and discuss the difference between reversible and irreversible processes. Discuss the importance of the first and second laws of thermodynamics as they apply to various forms of energy conversion and thermal pollution.

▷ Describe the processes that take place in an ideal heat engine taken through a **Carnot cycle**.

▷ Calculate the efficiency of a Carnot engine, and note that the efficiency of real heat engines is always less than the Carnot efficiency.

▷ Calculate entropy changes for reversible processes (such as one involving an ideal gas).

▷ Calculate entropy changes for irreversible processes, recognizing that the entropy change for an irreversible process is equivalent to that of a reversible process between the same two equilibrium states.

ANSWERS TO SELECTED CONCEPTUAL QUESTIONS

1.   What are some factors that affect the efficiency of automobile engines?

Answer   The most basic limit of efficiency for automobile engines is the Carnot cycle; the efficiency cannot **ever** exceed that of this ideal model. The efficiency is therefore affected by the maximum temperature that the engine block can withstand, and the intake and exhaust pressures. However, the Carnot cycle assumes that the exhaust gases can be released over an infinite period of time. If you want your engine to act more quickly than "never", you must accept additional losses. Other limits on efficiency are imposed by friction, the mass and acceleration of the engine parts, and the timing of the ignition.

□   □   □   □

2.   A steam-driven turbine is one major component of an electric power plant. Why is it advantageous to have the temperature of the steam as high as possible?

Answer   The most optimistic limit of efficiency in a steam engine is the ideal (Carnot) efficiency. This can be calculated from the high and low temperatures of the steam, as it expands:

$$e_c = \frac{T_h - T_c}{T_h} = 1 - \frac{T_c}{T_h}$$

The engine will be most efficient when the low temperature is extremely low, and the high temperature is extremely high. However, since the electric power plant is typically placed on the surface of the earth, there is a limit to how much the steam can expand, and how low the lower temperature can be. The only way to further increase the efficiency, then, is to raise the temperature of the hot steam, as high as possible.

□   □   □   □

4.   Discuss the change in entropy of a gas that expands (a) at constant temperature and (b) adiabatically.

Answer   (a) The expanding gas is doing work. If it is ideal, its constant temperature implies constant internal energy, and it must be taking in energy by heat equal in amount to the work it is doing. As energy enters the gas by heat its entropy increases. The derivation of equation 18.14 shows that the change in entropy is $\Delta S = nR\ln\left(V_f/V_i\right)$

(b) In a reversible adiabatic expansion there is no entropy change. We can say this is because the heat input is zero, or we can say it is because the temperature drops to compensate for the volume increase. In an irreversible adiabatic expansion the entropy increases. Equation 18.14 shows the entropy change in a free expansion.

□   □   □   □

18. The device shown in Figure Q18.18, called a thermoelectric converter, uses a series of semiconductor cells to convert internal energy to electric energy. In the picture at the left, both legs of the device are at the same temperature, and no electrical energy is produced. When one leg is at a higher temperature than the other, however, as in the picture on the right, electric energy is produced as the device extracts energy from the hot reservoir and drives a small electric motor. (a) Why does the temperature differential produce electric energy in this demonstration? (b) In what sense does this intriguing experiment demonstrate the second law of thermodynamics?

Figure Q18.18

Answer   (a) The semiconductor converter operates essentially like a thermocouple, which is a pair of wires of different metals, with a junction at each end. When the junctions are at different temperatures, a small voltage appears around the loop, so that the device can be used to measure temperature or (here) to drive a small motor.

(b) The second law states that an engine operating in a cycle cannot absorb energy by heat from one reservoir and expel it entirely by work. This exactly describes the first situation, where both legs are in contact with a single reservoir, and the thermocouple fails to produce work. To expel energy by work, the device must transfer energy by heat from a hot reservoir to a cold reservoir, as in the second situation.

SOLUTIONS TO SELECTED END-OF-CHAPTER PROBLEMS

3. A particular heat engine has a useful power output of 5.00 kW and an efficiency of 25.0%. The engine expels 8000 J of energy in each cycle. Find (a) the energy absorbed in each cycle and (b) the time for each cycle.

Solution   We are given that $|Q_c| = 8000 \text{ J}$

(a)   We have

$$e = \frac{W_{eng}}{|Q_h|} = \frac{|Q_h| - |Q_c|}{|Q_h|} = 1 - \frac{|Q_c|}{|Q_h|} = 0.250$$

   Isolating $|Q_h|$, we have

$$|Q_h| = \frac{|Q_c|}{1 - e} = \frac{8000 \text{ J}}{1 - 0.250} = 10.7 \text{ kJ} \qquad \Diamond$$

(b)   The work per cycle is

$$W_{eng} = |Q_h| - |Q_c| = 2667 \text{ J}$$

   From $\mathcal{P} = \dfrac{W_{eng}}{\Delta t}$,

$$\Delta t = \frac{W_{eng}}{\mathcal{P}} = \frac{2667 \text{ J}}{5000 \text{ J/s}} = 0.533 \text{ s} \qquad \Diamond$$

---

5. One of the most efficient engines ever built operates between 430 °C and 1870 °C. (a) What is its maximum theoretical efficiency? (b) The actual efficiency of the engine is 42.0%. How much useful power does the engine deliver if it absorbs 1.40×10^5 J of energy each second from its hot reservoir?

Solution   The engine is a steam turbine in an electric generating station:

$$T_c = 430 \text{ °C} = 703 \text{ K} \quad \text{and} \qquad T_h = 1870 \text{ °C} = 2143 \text{ K}$$

(a) $e_c = \dfrac{\Delta T}{T_h} = \dfrac{1440 \text{ K}}{2143 \text{ K}} = 0.672$ or   67.2% $\Diamond$

(b) $|Q_h| = 1.40 \times 10^5 \text{ J}$ $W_{eng} = 0.420|Q_h| = 5.88 \times 10^4 \text{ J}$

$$\mathcal{P} = \frac{W_{eng}}{\Delta t} = \frac{5.88 \times 10^4 \text{ J}}{1 \text{ s}} = 58.8 \text{ kW} \qquad \Diamond$$

---

9. An ideal gas is taken through a Carnot cycle. The isothermal expansion occurs at 250 °C, and the isothermal compression takes place at 50.0 °C. The gas absorbs 1200 J of energy from the hot reservoir during the isothermal expansion. Find (a) the energy expelled to the cold reservoir in each cycle and (b) the net work done by the gas in each cycle.

Solution

(a) For a Carnot cycle,

$$e_c = 1 - \frac{T_c}{T_h}$$

For any engine,

$$e = \frac{W_{eng}}{|Q_h|} = 1 - \frac{|Q_c|}{|Q_h|}$$

Therefore, for a Carnot engine,

$$1 - \frac{T_c}{T_h} = 1 - \frac{|Q_c|}{|Q_h|}$$

Then, since $|Q_c| = |Q_h| \left(\dfrac{T_c}{T_h} \right)$

$$|Q_c| = (1200 \text{ J}) \left(\frac{323 \text{ K}}{523 \text{ K}} \right) = 741 \text{ J} \qquad \Diamond$$

(b) The work is calculated as

$$W_{eng} = |Q_h| - |Q_c| = 1200 \text{ J} - 741 \text{ J} = 459 \text{ J} \qquad \Diamond$$

---

17. An ideal refrigerator or ideal heat pump is equivalent to a Carnot engine running in reverse. That is, energy Q_c is absorbed from a cold reservoir, and energy $|Q_h|$ is rejected to a hot reservoir. (a) Show that the work that must be supplied to run the refrigerator or pump is $W = Q_c(T_h - T_c)/T_c$. (b) Show that the coefficient of performance of the ideal refrigerator is $\text{COP} = T_c/(T_h - T_c)$.

Solution

(a) For a complete cycle $\Delta E_{int} = 0$, and

$$W = |Q_h| - |Q_c| = |Q_c| \left(\frac{|Q_h|}{|Q_c|} - 1 \right)$$

Since, for a Carnot cycle (and only for a Carnot cycle) $|Q_h|/|Q_c| = T_h/T_c$

Then,

$$W = |Q_c|(T_h - T_c)/T_c \qquad \Diamond$$

(b) From Equation 18.6, $\text{COP} = \dfrac{|Q_c|}{W}$, so

$$\text{COP} = \frac{T_c}{T_h - T_c} \qquad \Diamond$$

---

19. How much work does an ideal Carnot refrigerator require to remove 1.00 J of energy from helium at 4.00 K and reject this energy to a room-temperature (293-K) environment?

Solution $(COP)_{Carnot, refrig} = \dfrac{T_c}{\Delta T} = \dfrac{4.00 \text{ K}}{293 \text{ K} - 4.00 \text{ K}} = 0.0138 = \dfrac{|Q_c|}{W}$

$$W = \dfrac{|Q_c|}{COP} = \dfrac{1.00 \text{ J}}{0.0138} = 72.2 \text{ J} \qquad \lozenge$$

---

21. Calculate the change in entropy of 250 g of water heated slowly from 20.0 °C to 80.0 °C. (**Hint:** Note that $dQ = mc\,dT$.)

Solution We use the energy version of the nonisolated system model. To do the heating reversibly, put the water pot successively into contact with reservoirs at temperatures 20.0 °C + δ, 20.0 °C + 2δ, ... 80.0 °C, where δ is some small increment.

Then $\qquad \Delta S = \displaystyle\int_i^f \dfrac{dQ}{T} = \int_{T_i}^{T_f} mc\dfrac{dT}{T}$

Here T means the absolute temperature. We would ordinarily think of dT as the change in the Celsius temperature, but one Celsius degree of temperature change is the same size as one Kelvin of change, so dT is also the change in absolute T.

Then $\qquad \Delta S = mc \ln T \Big|_{T_i}^{T_f} = mc \ln\left(\dfrac{T_f}{T_i}\right)$

$$\Delta S = (0.250 \text{ kg})(4186 \text{ J/kg} \cdot \text{K}) \ln\left(\dfrac{353 \text{ K}}{293 \text{ K}}\right) = 195 \text{ J/K} \qquad \lozenge$$

---

25. A bag contains 50 red marbles and 50 green marbles. (a) You draw a marble at random from the bag, notice its color, return it to the bag, and repeat for a total of three draws. You record the result as the number of red marbles and the number of green marbles in the set of three. Construct a table listing each possible macrostate and the number of microstates within it. For example, RRG, RGR, and GRR are three microstates constituting the macrostate 1G, 2R. (b) Construct a table for the case in which you draw five marbles rather than three.

Solution: The combinations are as follows:

(a)

Result	Possible combinations	Total
All Red	RRR	1
2R, 1G	RRG, RGR, GRR	3
1R, 2G	RGG, GRG, GGR	3
All Green	GGG	1	◊

(b)

Result	Possible combinations	Total
All Red	RRRRR	1
4R, 1G	RRRRG, RRRGR, RRGRR, RGRRR, GRRRR	5
3R, 2G	RRRGG, RRGRG, RGRRG, GRRRG, RRGGR, RGRGR, GRRGR, RGGRR, GRGRR, GGRRR	10
2R, 3G	GGGRR, GGRGR, GRGGR, RGGGR, GGRRG, GRGRG, RGGRG, GRRGG, RGRGG, RRGGG	10
1R, 4G	GGGGR, GGGRG, GGRGG, GRGGG, RGGGG	5
All Green	GGGGG	1	◊

---

29. A 1500-kg car is moving at 20.0 m/s. The driver brakes to a stop. The brakes cool off to the temperature of the surrounding air, which is nearly constant at 20.0 °C. What is the total entropy change?

Solution The original kinetic energy of the car,

$$K = \tfrac{1}{2}mv^2 = \tfrac{1}{2}(1500 \text{ kg})(20.0 \text{ m/s})^2 = 300 \text{ kJ}$$

becomes irreversibly 300 kJ of extra internal energy in the brakes, the car, and its surroundings. Since their total heat capacity is so large, their equilibrium temperature will be approximately 20.0 °C. To carry them reversibly to this same final state, imagine putting 300 kJ into the car and its environment from a heater at 20.001°C. Then

$$\Delta S = \int_i^f \frac{dQ}{T} = \frac{1}{T}\int dQ_r = \frac{Q}{T} = \frac{300 \text{ kJ}}{293 \text{ K}}$$

$$\Delta S = 1.02 \text{ kJ/K} \qquad ◊$$

39. A house loses energy by conduction through the exterior walls and roof at a rate of 5000 J/s = 5.00 kW when the interior temperature is 22.0 °C and the outside temperature is –5.00 °C. Calculate the electric power required to maintain the interior temperature at 22.0 °C for the following two cases: (a) The electric power is used in electric resistance heaters (which convert all of the electric energy supplied into internal energy). (b) The electric power is used to drive an electric motor that operates the compressor of a heat pump that has a coefficient of performance equal to 60.0% of the Carnot-cycle value.

Solution

The electric heater should be 100% efficient, so $\mathcal{P} = 5$ kW in part (a). It sounds as if the heat pump is only 60% efficient, so we might expect $\mathcal{P} \cong 9$ kW in part (b).

Power is the amount of energy transferred per unit of time, so we can find the power in each case by examining the energy input as heat required for the house as a nonisolated system in steady state.

(a) We know that $\mathcal{P}_{electric} = H_{ET} / \Delta t$, so if all of the energy transferred into the heater by electrical transmission is stored in the heater as internal energy, then

$$\mathcal{P}_{electric} = \frac{H_{ET}}{\Delta t} = 5.00 \text{ kW} \qquad \lozenge$$

(b) For a heat pump, $\quad (\text{COP})_{Carnot} = \dfrac{T_h}{\Delta T} = \dfrac{295 \text{ K}}{27.0 \text{ K}} = 10.92$

$$\text{Actual COP} = (0.600)(10.92) = \frac{|Q_h|}{W} = \frac{|Q_h|/\Delta t}{W/\Delta t}$$

Therefore, to bring 5000 W of heat into the house only requires input power

$$\mathcal{P}_{heat\ pump} = \frac{W}{\Delta t} = \frac{|Q_h|/\Delta t}{\text{COP}} = \frac{5000 \text{ W}}{6.56} = 763 \text{ W} \qquad \lozenge$$

The result for the electric heater's power is consistent with our prediction, but the heat pump actually requires far **less** power than we expected. Since both types of heaters use electricity to operate, we can now see why it is more cost effective to use a heat pump even though it is less than 100% efficient!

41. In 1827, Robert Sterling, a Scottish minister, invented the **Sterling engine**, which has found a variety of applications ever since. Fuel is burned externally to warm one of the engine's two cylinders. A fixed quantity of inert gas moves cyclically between the cylinders, expanding in the hot one and contracting in the cold one. Figure P18.41 represents a model for its thermodynamic cycle. Consider n mol of an ideal monatomic gas being taken once through the cycle, consisting of two isothermal processes at temperatures $3T_i$ and T_i and two constant-volume processes. Determine, in terms of n, R, and T_i, (a) the net energy transferred by heat to the gas and (b) the efficiency of the engine.

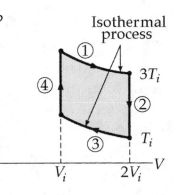

Figure P18.41

Solution   The Sterling engine need have no material exhaust, but it has energy exhaust.

(a)   For an isothermal process,

$$Q = nRT \ln\left(\frac{V_f}{V_i}\right)$$

Therefore, $Q_1 = nR(3T_i)\ln 2$ and $Q_3 = nRT_i \ln\left(\tfrac{1}{2}\right)$

The internal energy of a monatomic ideal gas is, from equation 16.18, $E_{int} = \frac{3}{2}nRT$. In the constant-volume processes,

$$Q_2 = \Delta E_{int,2} = \frac{3}{2}nR(T_i - 3T_i) \qquad \text{and} \qquad Q_4 = \Delta E_{int,4} = \frac{3}{2}nR(3T_i - T_i)$$

The net energy transferred by heat is then

$$Q = Q_1 + Q_2 + Q_3 + Q_4 \qquad \text{or} \qquad Q = 2nRT_i \ln 2 \qquad\qquad \lozenge$$

(b) $\left|Q_h\right|$ is the sum of the positive contributions to Q.

$$\left|Q_h\right| = Q_1 + Q_4 = 3nRT_i(1 + \ln 2)$$

Since the change in temperature for the complete cycle is zero, $\Delta E_{int} = 0$ and $W_{eng} = Q$.

Therefore, the efficiency is $e = \dfrac{W_{eng}}{\left|Q_h\right|} = \dfrac{Q}{\left|Q_h\right|} = \dfrac{2\ln 2}{3(1 + \ln 2)} = 0.273 = 27.3\%$ $\lozenge$

49. One mole of a monatomic ideal gas is taken through the cycle shown in Figure P18.49. At point A, the pressure, volume, and temperature are P_i, V_i, and T_i, respectively. In terms of R and T_i, find (a) the total energy entering the system by heat per cycle, (b) the total energy leaving the system by heat per cycle, (c) the efficiency of an engine operating in this cycle, and (d) the efficiency of an engine operating in a Carnot cycle between the same temperature extremes.

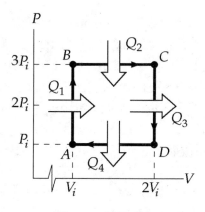

Figure P18.49

Solution   At point A, $P_iV_i = nRT_i$ with $n = 1.00$ mol

At point B, $3P_iV_i = nRT_B$ so $T_B = 3T_i$

At point C, $(3P_i)(2V_i) = nRT_C$ and $T_C = 6T_i$

At point D, $P_i(2V_i) = nRT_D$ and $T_D = 2T_i$

We find the energy transfer by heat for each step in the cycle using

$$C_V = \frac{3}{2}R \qquad \text{and} \qquad C_P = \frac{5}{2}R$$

$$Q_1 = Q_{AB} = C_V(3T_i - T_i) = 3RT_i \qquad Q_2 = Q_{BC} = C_P(6T_i - 3T_i) = 7.5RT_i$$

$$Q_3 = Q_{CD} = C_V(2T_i - 6T_i) = -6RT_i \qquad Q_4 = Q_{DA} = C_P(T_i - 2T_i) = -2.5RT_i$$

Therefore   (a) $Q_{in} = |Q_h| = Q_{AB} + Q_{DA} = 10.5RT_i$ ◊

(b) $Q_{out} = |Q_c| = |Q_{CD} + Q_{DA}| = 8.5RT_i$ ◊

(c) $e = \dfrac{|Q_h| - |Q_c|}{|Q_h|} = 0.190 = 19\%$ ◊

(d)   Carnot Efficiency, $e_c = 1 - \dfrac{T_c}{T_h} = 1 - \dfrac{T_i}{6T_i} = 0.833 = 83.3\%$ ◊

53. A system consisting of n mol of an ideal gas undergoes two reversible processes. It starts with pressure P_i and volume V_i, expands isothermally, and then contracts adiabatically to reach a final state with pressure P_i and volume $3V_i$. (a) Find its change in entropy in the isothermal process. Note that no entropy change occurs in the adiabatic process. (b) Explain why the answer to part (a) must be the same as the answer to Problem 52.

Solution   The diagram shows the isobaric process considered in Problem 52 as AB. The processes considered in this problem are AC and CB.

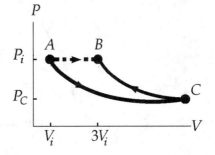

(a)   For isotherm (AC), $$P_A V_A = P_C V_C$$

For adiabat (CB), $$P_C V_C{}^\gamma = P_B V_B{}^\gamma$$

Combining these gives

$$V_C = \left(\frac{P_B V_B{}^\gamma}{P_A V_A} \right)^{1/(\gamma-1)} = \left[\left(\frac{P_i}{P_i} \right) \frac{(3V_i)^\gamma}{V_i} \right]^{1/(\gamma-1)} = \left(3^{\gamma/(\gamma-1)} \right) V_i$$

Therefore,

$$\Delta S_{AC} = nR \ln \left(\frac{V_C}{V_A} \right) = nR \ln \left[3^{\gamma/(\gamma-1)} \right] = \frac{nR\gamma \ln 3}{\gamma - 1} \qquad \Diamond$$

(b)   Since the change in entropy is path independent, $\Delta S_{AB} = \Delta S_{AC} + \Delta S_{CB}$

But because (CB) is adiabatic, $\Delta S_{CB} = 0$

Then $\Delta S_{AB} = \Delta S_{AC}$

The answer to problem 52 was stated as $\Delta S_{AB} = nC_P \ln 3$

Because $\gamma = C_P/C_V$, $C_V = C_P/\gamma$ and $C_P - C_V = R$

gives $C_P - C_P/\gamma = R$ so $\gamma C_P - C_P = \gamma R$ and $C_P = \gamma R/(\gamma-1)$

Thus, the answers to problems 52 and 53 are in fact equal. $\Diamond$

Chapter 19

Electric Forces and Electric Fields

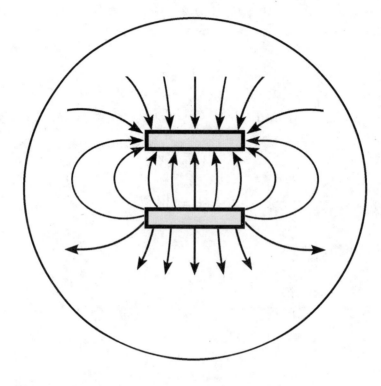

INTRODUCTION

The electromagnetic force between charged particles is one of the fundamental forces of nature. In this chapter, we begin by describing some of the basic properties of electric forces. We then discuss Coulomb's law, which is the fundamental law of force between any two stationary charged particles. The concept of an electric field associated with a charge distribution is then introduced, and the interaction of the field with other charges is described. The method of using Coulomb's law to calculate electric fields of a given charge distribution is discussed, and several examples are given.

Gauss's law is presented as an alternative procedure for calculating electric fields due to symmetric charge distributions. This formulation is based on the fact that the fundamental electrostatic force between point charges is an inverse-square law. Although Gauss's law is a consequence of Coulomb's law, Gauss's law is much more convenient for calculating the electric field of highly symmetric charge distributions. Furthermore, Gauss's law serves as a guide for understanding more complicated problems.

NOTES FROM SELECTED CHAPTER SECTIONS

19.2 Properties of Electric Charges

Electric charge has the following important properties:
- There are two kinds of charges in nature, positive and negative, with the property that unlike charges attract one another and like charges repel one another.
- Charge is conserved.
- Charge is quantized.

19.3 Insulators and Conductors

Conductors are materials in which electric charges move freely under the influence of an electric field; **insulators** are materials that do not readily transport charge.

19.4 Coulomb's Law

Experiments show that an **electric force** between a pair of point charges is:
- of equal magnitude and opposite direction on the two charges.
- attractive if the charges are of opposite sign and repulsive if the charges have the same sign.
- inversely proportional to the square of the separation, r, between the two charges and is along the line joining them.
- proportional to the product of the magnitudes of the charges.

19.5 Electric Fields

An electric field exists at some point if a test charge at rest placed at that point experiences an electrical force.

The electric field vector **E** at some point in space is defined as the electric force **F** acting on a positive test charge placed at that point divided by the magnitude of the test charge q_0.

At any point the total electric field created by a group of discrete point charges equals the vector sum of the electric fields due to each of the charges individually.

19.6   Electric Field Lines

A convenient aid for visualizing electric field patterns is to draw lines pointing in the same direction as the electric field vector at any point. These lines, called electric field lines, are related to the electric field in any region of space in the following manner:
- The electric field vector $\mathbf{E}$ is **tangent** to the electric field line at each point.
- The number of lines per unit area through a surface perpendicular to the lines is proportional to the strength of the electric field in that region. Thus, $\mathbf{E}$ is large when the field lines are close together and small when they are far apart.

The rules for drawing electric field lines for any charge distribution are as follows:
- The lines must begin on positive charges and terminate on negative charges, or at infinity in the case of an excess of charge.
- The number of lines drawn leaving a positive charge or approaching a negative charge is proportional to the magnitude of the charge.
- No two field lines can cross.

19.9   Gauss's Law

Gauss's law states that the net electric flux through a closed gaussian surface is equal to the net charge inside the surface divided by ϵ_0.

When using Gauss's law to calculate an electric field, choose the gaussian surface so that it has the same symmetry as the charge distribution.

19.11   Conductors in Electrostatic Equilibrium

A conductor in electrostatic equilibrium has the following properties:
- The electric field is zero everywhere inside the conductor.
- Any excess charge on an isolated conductor resides entirely on its surface.
- The electric field just outside a charged conductor is perpendicular to the conductor's surface and has a magnitude σ/ϵ_0, where σ is the charge per unit area at that point.
- On an irregularly shaped conductor, charge tends to accumulate at locations where the radius of curvature of the surface is the smallest, that is, at sharp points.

EQUATIONS AND CONCEPTS

The **magnitude** of the **electrostatic force** between two stationary point charges, q_1 and q_2, separated by a distance r is given by **Coulomb's law**.

$$F_e = k_e \frac{|q_1||q_2|}{r^2}$$

where $\quad k_e = \dfrac{1}{4\pi\epsilon_0}$

$$\epsilon_0 = 8.8542 \times 10^{-12} \ C^2/N \cdot m^2 \qquad (19.1)$$

In calculations, an approximate value for k_e may be used.

$$k_e = 8.99 \times 10^9 \ N \cdot m^2/C^2$$

The direction of the electrostatic force on each charge is determined from the experimental observation that like sign charges experience forces of mutual repulsion and unlike sign charges attract each other. By virtue of Newton's third law, the magnitude of the force on each of the two charges is the same regardless of the relative magnitude of the values of q_1 and q_2.

The electric force between two charges can be expressed in vector form. $\mathbf{F}_{12}$ is the force **on** q_2 **due to** q_1 and $\hat{\mathbf{r}}_{12}$ is a unit vector directed from q_1 to q_2. Coulomb's law applies exactly to point charges or particles.

$$\mathbf{F}_{12} = k_e \frac{q_1 q_2}{r^2} \hat{\mathbf{r}}_{12} \qquad (19.2)$$

In cases where there are more than two charges present, the **resultant force** on any one charge is the vector sum of the forces exerted on that charge by the remaining individual charges present. The **principle of superposition** applies.

$$\mathbf{F}_i = \sum_{j=1}^{N} \mathbf{F}_{ji}$$

The **electric field** at any point in space is defined as the ratio of electric force per charge exerted on a small positive test charge placed at the point where the field is to be determined.

$$\mathbf{E} \equiv \frac{\mathbf{F}_e}{q_0} \qquad (19.3)$$

The above definition (Equation 19.3) together with Coulomb's law leads to an expression for calculating the **electric field a distance** r **from a point charge,** q. In this case the unit vector $\hat{\mathbf{r}}$ is **directed away from** q **and toward the point** P **where the field is to be calculated**. The direction of the electric field is radially outward from a positive point charge and radially inward toward a negative point charge.

$$\mathbf{E} = k_e \frac{q}{r^2} \hat{\mathbf{r}} \qquad (19.5)$$

The superposition principle holds when the electric field at a point is due to a number of point charges.

$$\mathbf{E} = k_e \sum_i \frac{q_i}{r_i^2} \hat{\mathbf{r}}_i \qquad (19.6)$$
(vector sum)

Electric field lines are a convenient graphical representation of electric field patterns. These lines are drawn so that the electric field vector, **E**, is tangent to the electric field lines at each point. Also, the number of lines per unit area through a surface perpendicular to the lines is proportional to the strength or magnitude of the electric field over the region. In every case, electric field lines must begin on positive charges and terminate on negative charges or at infinity; the number of lines leaving or approaching a charge is proportional to the magnitude of the charge; and no two field lines can cross.

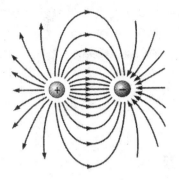

When the electric field is due to a **continuous charge distribution**, the contributions to the field by the elements of charge must be integrated over the total line, surface, or volume, which contains the charge.

$$\mathbf{E} = k_e \int\limits_{\substack{\text{All} \\ \text{Charge}}} \frac{dq}{r^2}\,\hat{\mathbf{r}} \qquad (19.7)$$

In order to perform the integration described above, it is convenient to represent a charge increment dq as the product of an element of length, area, or volume and the **charge density over that region**. **Note**: For those cases in which the charge is not uniformly distributed, the densities λ, σ, and ρ must be stated as functions of position.

For an element of length dx

$$dq = \lambda\,dx$$

For an element of area dA

$$dq = \sigma\,dA$$

For an element of volume dV

$$dq = \rho\,dV$$

For uniform charge distributions the volume charge density (ρ), the surface charge density (σ), and the linear charge density (λ) can be calculated from the total charge and the total volume, area, or length.

$$\rho \equiv \frac{Q}{V} \tag{19.8}$$

$$\sigma \equiv \frac{Q}{A} \tag{19.9}$$

$$\lambda \equiv \frac{Q}{\ell} \tag{19.10}$$

The **electric flux** is proportional to the number of electric field lines that penetrate some surface. For a **plane surface** in a **uniform field**, the flux depends on the angle between the normal to the surface and the direction of the field.

$$\Phi_E = EA\cos\theta \tag{19.18}$$

In the case of a **surface of arbitrary shape** in the region of a **nonuniform** field, the flux is calculated by integrating the normal component of the field over the surface in question.

$$\Phi_E = \int_{\text{surface}} \mathbf{E} \cdot d\mathbf{A} \tag{19.19}$$

Gauss's law states that when Φ_E (Eq. 19.19) is evaluated over a **closed** surface (gaussian surface), the result equals **the net charge enclosed by the surface** divided by the constant ϵ_0. The symbol $\oint$ in this equation indicates that the integral must be evaluated over a **closed** surface.

$$\Phi_E = \oint_{\substack{\text{Closed} \\ \text{surface}}} \mathbf{E} \cdot d\mathbf{A} = \frac{q_{in}}{\epsilon_0} \tag{19.22}$$

The electric field just outside the surface of a charged conductor in equilibrium can be expressed in terms of the surface charge density on the conductor. Just outside the conductor, the field is normal to the surface.

$$E_n = \frac{\sigma}{\epsilon_0} \tag{19.25}$$

SUGGESTIONS, SKILLS, AND STRATEGIES

Problem-Solving Hints for Electric Forces and Fields:

- **Units**: When performing calculations that involve the use of the Coulomb constant k_e that appears in Coulomb's law, charges must be in coulombs and distances in meters. If they are given in other units, you must convert them to SI.

- **Applying Coulomb's law to point charges**: It is important to use the superposition principle properly when dealing with a collection of interacting point charges. When several charges are present, the resultant force on any one of them is found by finding the individual force that every other charge exerts on it and then finding the vector sum of all these forces. The magnitude of the force that any charged object exerts on another is given by Coulomb's law, and the direction of the force is found by noting that the forces are repulsive between like charges and attractive between unlike charges.

- **Calculating the electric field due to a group of point charges**: Remember that the superposition principle can also be applied to electric fields, which are also vector quantities. To find the total electric field at a given point, first calculate the electric field at the point due to each individual charge. The resultant field at the point is the vector sum of the fields due to the individual charges.

- **Calculating the electric field due to a continuous charge distribution**: To evaluate the electric field of a continuous charge distribution, it is convenient to employ the concept of charge density. Charge density can be written in different ways: charge per unit volume, ρ; charge per unit area, σ; or charge per unit length, λ. The total charge distribution is then subdivided into a small element of volume dV, area dA, or length dx. Each element contains an increment of charge dq (equal to ρdV, σdA, or λdx). If the charge is **nonuniformly** distributed over the region, then the charge densities must be written as functions of position. For example, if the charge density along a line or long bar is proportional to the distance from one end of the bar, then the linear charge density could be written as $\lambda = bx$ and the charge increment dq becomes $dq = bx\,dx$.

- **Symmetry**: Whenever dealing with either a distribution of point charges or a continuous charge distribution, take advantage of any symmetry in the system to simplify your calculations.

Problem-Solving Hints for Gauss's Law

Gauss's law is a very powerful theorem, which relates any charge distribution to the resulting electric field at any point in the vicinity of the charge. In this chapter you should learn how to apply Gauss's law to those cases in which the charge distribution has a sufficiently high degree of symmetry. As you review Examples 19.9 through 19.12 of the text, observe how each of the following steps has been included in the application of the equation $\oint \mathbf{E} \cdot d\mathbf{A} = q/\epsilon_0$ to that particular situation.

- The gaussian surface should be chosen to have the **same symmetry as the charge distribution.**

- The dimensions of the surface must be such that the surface includes the point where the electric field is to be calculated.

- From the symmetry of the charge distribution, you should be able to correctly describe the direction of the electric field vector, $\mathbf{E}$, relative to the direction of an element of surface area vector, $d\mathbf{A}$, over each region of the gaussian surface.

- From the symmetry of the charge distribution, you should also be able to identify one or more portions of the closed surface (and in some cases the entire surface) over which the magnitude of $\mathbf{E}$ remains constant.

- Write $\mathbf{E} \cdot d\mathbf{A}$ as $E \, dA \cos\theta$, and divide the surface into separate regions such that over each region:

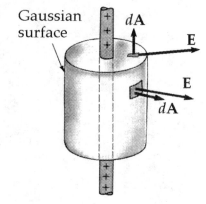

Gaussian surface

$E(dA)(\cos\theta)$ will equal 0 when $\mathbf{E} \perp d\mathbf{A}$ or when $\mathbf{E} = 0$. An example of the first case may be seen at the top of the accompanying figure; the second case will typically be true of the area inside the surface of a conductor.

$E(dA)(\cos\theta)$ will equal $E \, dA$ when $\mathbf{E} \parallel d\mathbf{A}$. An example of this is shown in the figure to the right, over the curved portion of the cylindrical gaussian surface.

$E(dA)(\cos\theta) = -E(dA)$ when $\mathbf{E}$ and $d\mathbf{A}$ are oppositely directed.

- If the gaussian surface has been chosen and subdivided so that the magnitude of **E** is **constant** over those regions where **E**·d**A** = $E\,dA$, then over each of those regions

$$\int \mathbf{E} \cdot d\mathbf{A} = E \int dA = E \text{ (area of region)}$$

- The total charge enclosed by the gaussian surface is that portion of the charge **inside** the gaussian surface.

- Once the left and right sides of Gauss's law have been evaluated, you can calculate the electric field on the gaussian surface, assuming the charge distribution is given in the problem. Conversely, if the electric field is known, you can calculate the charge distribution that produces the field.

REVIEW CHECKLIST

▷ Describe the fundamental properties of electric charge and the nature of electrostatic forces between charged bodies.

▷ Use Coulomb's law to determine the net electrostatic force on a point electric charge due to a known distribution of a finite number of point charges.

▷ Calculate the electric field **E** (magnitude and direction) at a specified location in the vicinity of a group of point charges.

▷ Calculate the electric field due to a continuous charge distribution. The charge may be distributed uniformly or nonuniformly along a line, over a surface, or throughout a volume.

▷ Calculate the **electric** flux through a surface; in particular, find the net electric flux through a **closed** surface.

▷ Understand that a Gaussian surface is a **closed** mathematical surface; the surface may be on or within a conductor, an insulator, or in space. Also remember that the net electric flux through a closed gaussian surface is equal to the net charge enclosed by the surface divided by the constant ϵ_0.

▷ Use Gauss's law to evaluate the electric field at points in the vicinity of charge distributions, which exhibit spherical, cylindrical, or planar symmetry.

ANSWERS TO SELECTED CONCEPTUAL QUESTIONS

3. A balloon is negatively charged by rubbing and then clings to a wall. Does this mean that the wall is positively charged? Why does the balloon eventually fall?

Answer

No. The balloon induces polarization of the molecules in the wall, so that a layer of positive charge exists near the balloon. This is just like the situation in Figure 19.6a, except that the signs of the charges are reversed. The attraction between these charges and the negative charges on the balloon is stronger than the repulsion between the negative charges on the balloon and the negative charges in the polarized molecules (because they are farther from the balloon), so that there is a net attractive force toward the wall. Ionization processes in the air surrounding the balloon provide ions to which excess electrons in the balloon can transfer, reducing the charge on the balloon and eventually causing the attractive force to be insufficient to support the weight of the balloon.

6. Would life be different if the electron were positively charged and the proton were negatively charged? Does the choice of signs have any bearing on physical and chemical interactions? Explain.

Answer

No, life would not be different. The character and effect of electric forces is defined by (1) the fact that there are only two types of electric charge — positive and negative, and (2) the fact that opposite charges attract, while like charges repel. The choice of signs is completely arbitrary.

As a related exercise, you might consider what would happen in a world where there were three types of electric charge, or in a world where opposite charges repelled, and like charges attracted.

15. If the total charge inside a closed surface is known but the distribution of the charge is unspecified, can you use Gauss's law to find the electric field? Explain.

Answer

No. If we wish to use Gauss's law to find the electric field, we must be able to bring the electric field, **E**, out of the integral. This can be done, in some cases — when the field is constant, for example. However, since we do not know the charge distribution, we cannot claim that the field is constant, and thus cannot find the electric field.

To illustrate this point, consider a sphere that contained a net charge of 100 μC. The charges could be located near the center, or they could all be grouped at the northernmost point within the sphere. In either case, the net electric flux would be the same, but the electric field would vary greatly.

17. A common demonstration involves charging a balloon, which is an insulator, by rubbing it on your head, and touching the balloon to a ceiling or wall, which is also an insulator. The electrical attraction between the charged balloon and the neutral wall results in the balloon sticking to the wall. Imagine now that we have two infinitely large flat sheets of insulating material. One is charged and the other is neutral. If these are brought into contact, will an attractive force exist between them, as there was for the balloon and the wall?

Answer

There will not be an attractive force. There are two factors to consider in the attractive force between a balloon and a wall, or between any pair of charged and neutral objects. The first factor is that the molecules in the wall will orient themselves with their negative ends toward the balloon, and their positive ends pointing away from the balloon. The second factor to consider is that the balloon is of finite curved dimensions, and thus the molecules in the wall are in a nonuniform electric field. Therefore the nearby "negative ends" of the molecules in the wall will experience an attractive electrostatic force that will be greater in magnitude than the repulsive force exerted on the more distant 'positive ends" of the molecules. The net result is an overall force of attraction.

Now consider the infinite sheets brought into contact. The polarization of the molecules in the neutral sheet will indeed occur, as in the wall. But the electric field from the charged sheet is **uniform**, and therefore is independent of the distance from the sheet. Thus, both the negative and positive charges in the neutral sheet will experience the same electric field and the same magnitude of electric force. The attractive force on the negative charges will cancel with the repulsive force on the positive charges, and there will be no net force.

□　□　□　□

20. A person is placed in a large hollow metallic sphere that is insulated from ground. If a large charge is placed on the sphere, will the person be harmed on touching the inside of the sphere? Explain what will happen if the person also has an initial charge whose sign is opposite that of the charge on the sphere.

Answer

The metallic sphere is a good conductor, so any excess charge on the sphere will reside on the outside of the sphere. From Gauss's law, we know that the field inside the sphere will then be zero. As a result, when the person touches the inside of the sphere, no charge will be exchanged between the person and the sphere, and the person will not be harmed.

What happens, then, if the person has an initial charge? Regardless of the sign of the person's initial charge, the charges in the conducting surface will redistribute themselves to maintain a net zero charge within the **conducting metal**. Thus, if the person has a 5.00 μC charge on his skin, exactly –5.00 μC will gather on the inner surface of the sphere, so that the electric field inside the metal will be zero. When the person touches the metallic sphere then, he will receive a shock due to the charge on his own skin.

□　□　□　□

SOLUTIONS TO SELECTED END-OF-CHAPTER PROBLEMS

3. The Nobel laureate in physics Richard Feynman once said that if two persons stood at arm's length from each other and each person had 1% more electrons than protons, the force of repulsion between them would be enough to lift a "weight" equal to that of the entire Earth. Carry out an order-of-magnitude calculation to substantiate this assertion.

Solution

Suppose each person has mass 70 kg. In terms of elementary charges, each person consists of precisely equal numbers of protons and electrons and a nearly equal number of neutrons. The electrons comprise very little of the mass, so we find the number of protons-and-neutrons in each person:

$$(70 \text{ kg})\left(\frac{1 \text{ u}}{1.66 \times 10^{-27} \text{ kg}}\right) = 4 \times 10^{28} \text{ u}$$

Of these, nearly one half, 2×10^{28}, are protons, and 1% of this is 2×10^{26}, constituting a charge of $(2 \times 10^{26})(1.60 \times 10^{-19} \text{ C}) = 3 \times 10^7$ C. Thus, Feynman's force is

$$F = \frac{k_e q_1 q_2}{r^2} = \frac{(8.99 \times 10^9 \text{ N} \cdot \text{m}^2 / \text{C}^2)(3 \times 10^7 \text{ C})^2}{(0.5 \text{ m})^2} \sim 10^{26} \text{ N}$$

where we have used a half-meter arm's length.

According to the particle in a gravitational field model, if the Earth were in an externally-produced uniform gravitational field of magnitude 9.80 m/s², it would weigh

$$F_g = mg = (6 \times 10^{24} \text{ kg})(10 \text{ m/s}^2) \sim 10^{26} \text{ N}$$

Thus, the forces are of the same order of magnitude. ◊

7. Three points charges are located at the corners of an equilateral triangle as shown is Figure P19.7. Calculate the net electric force on the 7.00‑μC charge.

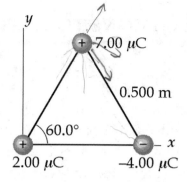

Solution The 7.00‑μC charge experiences a repulsive force $\mathbf{F}_1$ due to the 2.00‑μC charge, and an attractive force $\mathbf{F}_2$ due to the –4.00‑μC charge, where $F_2 = 2F_1$. If we sketch vectors representing $\mathbf{F}_1$ and $\mathbf{F}_2$ and the resultant, $\mathbf{F}$ (see figure at lower right) we find that the resultant appears to be about the same magnitude as F_2 and is directed to the right about 30.0° below the horizontal.

Figure P19.7

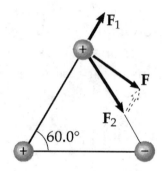

We can find the net electric force by adding the two separate forces acting on the 7.00‑μC charge. These individual forces can be found by applying Coulomb's law to each pair of charges.

The force on the 7.00‑μC charge by the 2.00‑μC charge is

$$\mathbf{F}_1 = k_e \frac{q_1 q_2}{r^2}\hat{\mathbf{r}} = \frac{\left(8.99 \times 10^9 \ \text{N}\cdot\text{m}^2/\text{C}^2\right)\left(7.00 \times 10^{-6}\ \text{C}\right)\left(2.00 \times 10^{-6}\ \text{C}\right)}{\left(0.500\ \text{m}\right)^2}\left(\cos 60°\mathbf{i} + \sin 60°\mathbf{j}\right)$$

$$\mathbf{F}_1 = \left(0.252\,\mathbf{i} + 0.436\,\mathbf{j}\right)\ \text{N}$$

Similarly, the force on the 7.00‑μC charge by the –4.00‑μC charge is

$$\mathbf{F}_2 = k_e \frac{q_1 q_3}{r^2}\hat{\mathbf{r}} = -\frac{\left(8.99 \times 10^9 \ \text{N}\cdot\text{m}^2/\text{C}^2\right)\left(7.00 \times 10^{-6}\ \text{C}\right)\left(-4.00 \times 10^{-6}\ \text{C}\right)}{\left(0.500\ \text{m}\right)^2}\left(\cos 60°\mathbf{i} - \sin 60°\mathbf{j}\right)$$

$$\mathbf{F}_2 = \left(0.503\,\mathbf{i} - 0.872\,\mathbf{j}\right)\ \text{N}$$

Thus, the total force on the 7.00‑μC charge, expressed as a set of components, is

$$\mathbf{F} = \mathbf{F}_1 + \mathbf{F}_2 = \left(0.755\,\mathbf{i} - 0.436\,\mathbf{j}\right)\ \text{N}$$

We can also write the total force as:

$$F = \sqrt{(0.755 \text{ N})^2 + (0.436 \text{ N})^2} \qquad \text{at} \qquad \tan^{-1}\left(\frac{0.436 \text{ N}}{0.755 \text{ N}}\right) \text{ below the } +x \text{ axis}$$

$$\mathbf{F} = 0.872 \text{ N at } 30.0° \text{ below the } +x \text{ axis} \qquad\qquad \diamond$$

Our calculated answer agrees with our initial estimate. An equivalent approach to this problem would be to find the net electric field due to the two lower charges and apply $\mathbf{F} = q\mathbf{E}$ to find the force on the upper charge in this electric field.

---

17. A uniformly charged insulating rod of length 14.0 cm is bent into the shape of a semicircle, as shown in Figure P19.17. The rod has a total charge of –7.50 μC. Find the magnitude and direction of the electric field at O, the center of the semicircle.

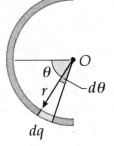

Solution   Let λ be the charge per unit length.

Then, $\qquad\qquad\qquad dq = \lambda\, ds = \lambda\, \theta d$

Figure P19.17
(modified)

and $\qquad\qquad\qquad dE = \dfrac{k_e dq}{r^2}$

In component form, $\qquad E_y = 0$ (from symmetry) $\qquad\qquad\qquad dE_x = dE\cos\theta$

Integrating, $\qquad E_x = \displaystyle\int dE_x = \int \frac{k_e \lambda r \cos\theta}{r^2}\, d\theta = \frac{k_e \lambda}{r} \int_{-\pi/2}^{\pi/2} \cos\theta\, d\theta = \frac{2k_e \lambda}{r}$

But $\qquad\qquad Q_{total} = \lambda\ell$, where $\ell = 0.140$ m, and $r = \ell/\pi$

Thus, $\qquad\qquad E_x = \dfrac{2\pi k_e Q}{\ell^2} = \dfrac{2\pi\left(8.99\times10^9 \text{ N}\cdot\text{m}^2/\text{C}^2\right)\left(-7.50\times10^{-6} \text{ C}\right)}{(0.140 \text{ m})^2}$

$$\mathbf{E} = \left(-2.16\times10^7 \text{ N/C}\right)\mathbf{i} \qquad\qquad\qquad \diamond$$

---

21. A negatively charged rod of finite length has a uniform charge per unit length. Sketch the electric field lines in a plane containing the rod.

Solution

Since the rod has negative charge, field lines point inwards. Any field line points nearly toward the center of the rod at large distances, where the rod would look like just a point charge. The lines curve to reach the rod perpendicular to its surface, where they end at equally-spaced points.

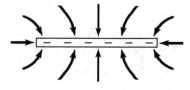

23. A proton accelerates from rest in a uniform electric field of 640 N/C. At some later time, its speed has reached 1.20×10^6 m/s (nonrelativistic, since v is much less than the speed of light). (a) Find the acceleration of the proton. (b) How long does it take the proton to reach this speed? (c) How far has it moved in this time? (d) What is its kinetic energy at this time?

Solution

(a) We use the particle in an electric field model and the particle under net force model.

$$a = \frac{F}{m} = \frac{qE}{m} = \frac{\left(1.602 \times 10^{-19} \text{ C}\right)\left(640 \text{ N/C}\right)}{1.67 \times 10^{-27} \text{ kg}} = 6.14 \times 10^{10} \text{ m/s}^2 \qquad \Diamond$$

(b) We use the particle under constant acceleration model.

$$\Delta t = \frac{\Delta v}{a} = \frac{1.20 \times 10^6 \text{ m/s}}{6.14 \times 10^{10} \text{ m/s}^2} = 19.5 \ \mu\text{s} \qquad \Diamond$$

(c) $\Delta x = v_i t + \frac{1}{2} a t^2 = 0 + \frac{1}{2}\left(6.14 \times 10^{10} \text{ m/s}^2\right)\left(19.5 \times 10^{-6} \text{ s}\right)^2 = 11.7 \text{ m} \qquad \Diamond$

(d) $K = \frac{1}{2} m v^2 = \frac{1}{2}\left(1.67 \times 10^{-27} \text{ kg}\right)\left(1.20 \times 10^6 \text{ m/s}\right)^2 = 1.20 \times 10^{-15} \text{ J} \qquad \Diamond$

29. A 40.0-cm-diameter loop is rotated in a uniform electric field until the position of maximum electric flux is found. The flux in this position is measured to be $5.20 \times 10^5 \ \mathrm{N \cdot m^2/C}$. What is the magnitude of the electric field?

Solution   We calculate the flux as

$$\Phi = \mathbf{E} \cdot \mathbf{A} = EA\cos\theta$$

The maximum value of the flux occurs when

$$\theta = 0$$

Therefore, we can calculate the field strength at this point as

$$E = \frac{\Phi_{max}}{A} = \frac{\Phi_{max}}{\pi r^2} \text{ of field}$$

$$E = \frac{5.20 \times 10^5}{\pi (0.200 \ \mathrm{m})^2} = 4.14 \times 10^6 \ \mathrm{N/C}$$ ◊

33. A point charge Q is located just above the center of the flat face of a hemisphere of radius R, as shown in Figure P19.33. What is the electric flux (a) through the curved surface and (b) through the flat face?

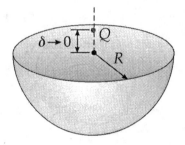

Figure P19.33

Solution   With δ very small, all points on the hemisphere are nearly at distance R from the charge, so the field everywhere on the curved surface is $k_e Q/R^2$ radially outward.

(a) The flux over the curved surface is this field strength times the area of half a sphere:

$$\Phi_{curved} = \left(k_e \frac{Q}{R^2}\right)\left(\tfrac{1}{2}\right)\left(4\pi R^2\right) = \frac{1}{4\pi\epsilon_0}Q(2\pi) = \frac{Q}{2\epsilon_0}$$ ◊

(b) The closed surface encloses zero charge, so Gauss's law gives:

$$\Phi_{curved} + \Phi_{flat} = 0$$ ◊

$$\Phi_{flat} = -\Phi_{curved} = \frac{-Q}{2\epsilon_0}$$ ◊

uniform

37. Consider a long cylindrical charge distribution of radius R with a uniform charge density ρ. Find the electric field at distance r from the axis where $r < R$.

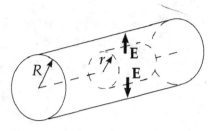

Solution

If ρ is positive, the field must everywhere be radially outward. Choose as the gaussian surface a cylinder of length L and radius r, contained inside the charged rod. Its volume is $\pi r^2 L$ and it encloses charge $\rho \pi r^2 L$. The circular end caps have no electric flux through them; there $\mathbf{E} \cdot d\mathbf{A} = 0$. The curved surface has $\mathbf{E} \cdot d\mathbf{A} = E \, dA \cos 0°$, and E must be the same strength everywhere over the curved surface.

Then $\oint \mathbf{E} \cdot d\mathbf{A} = \dfrac{q}{\epsilon_0}$ becomes

$$E \oint_{\substack{\text{Curved} \\ \text{surface}}} dA = \frac{\rho \pi r^2 L}{\epsilon_0}$$

Noting that $2\pi r L$ is the lateral surface area of the cylinder,

$$E(2\pi r L) = \frac{\rho \pi r^2 L}{\epsilon_0}$$

Thus, $\qquad \mathbf{E} = \dfrac{\rho r}{2\epsilon_0}$ radially away from the axis $\qquad\qquad\qquad$ ◊

---

43. A thin square conducting plate 50.0 cm on a side lies in the xy plane. If a total charge of 4.00×10^{-8} C is placed on the plate, find (a) the charge density on the plate, (b) the electric field just above the plate, and (c) the electric field just below the plate. You may assume that the charge density is uniform.

Solution   In this problem ignore "edge" effects and assume that the total charge distributes uniformly over each side of the plate (one half the total charge on each side).

(a) $\quad \sigma = \dfrac{q}{A} = \dfrac{4.00 \times 10^{-8} \text{ C}}{2(0.500 \text{ m})^2} = 8.00 \times 10^{-8} \text{ C/m}^2$ $\qquad\qquad$ ◊

(b) Just above the plate, $\quad E = \dfrac{\sigma}{\epsilon_0} = \dfrac{8.00 \times 10^{-8} \text{ C/m}^2}{8.85 \times 10^{-12} \text{ C}^2/\text{N} \cdot \text{m}^2} = 9.04 \times 10^3 \text{ N/C}$ upward $\qquad$ ◊

(c) Just below the plate, $\quad E = \dfrac{\sigma}{\epsilon_0} = 9.04 \times 10^3 \text{ N/C}$ downward $\qquad\qquad$ ◊

---

55. Two small spheres of mass m are suspended from strings of length ℓ that are connected at a common point. One sphere has charge Q; the other has charge $2Q$. Assume the angles θ_1 and θ_2 that the strings make with the vertical are small. (a) How are θ_1 and θ_2 related? (b) Show that the distance r between the spheres is given by

$$r \cong \left(\frac{4k_e Q^2 \ell}{mg} \right)^{1/3}$$

Solution   We use the particle in equilibrium model.

(a)   The spheres have different charges, but each exerts an equal force on the other, given by $F_e = k_e(Q)(2Q)/r^2$, where r is the distance between them. Since their masses are equal,

$$\theta_2 = \theta_1 \qquad \Diamond$$

(b)   For equilibrium, $\Sigma F_y = 0$: $\qquad T\cos\theta - mg = 0$

Thus $\qquad\qquad\qquad\qquad T = \dfrac{mg}{\cos\theta}$

$\Sigma F_x = 0$: $\qquad\qquad\qquad F_e - T\sin\theta = 0$

Substituting for T, $\qquad\quad F_e = \dfrac{mg\sin\theta}{\cos\theta} = mg\tan\theta$

For small angles, $\qquad\quad \tan\theta \cong \sin\theta = \dfrac{r}{2\ell}$

Therefore, $\qquad\qquad\qquad F_e \cong mg\dfrac{r}{2\ell}$

The force F_e is $\qquad\qquad \dfrac{k_e Q(2Q)}{r^2} \cong mg\dfrac{r}{2\ell}$

so that $\qquad\qquad\qquad 4k_e Q^2 \ell \cong mgr^3$

and $\qquad\qquad\qquad r \cong \left(\dfrac{4k_e Q^2 \ell}{mg} \right)^{1/3} \qquad \Diamond$

59. A solid, insulating sphere of radius a has a uniform charge density ρ and a total charge Q. Concentric with this sphere is an uncharged, conducting hollow sphere whose inner and outer radii are b and c, as shown in Figure P19.59. (a) Find the magnitude of the electric field in the regions $r < a$, $a < r < b$, $b < r < c$, and $r > c$. (b) Determine the induced charge per unit area on the inner and outer surfaces of the hollow sphere.

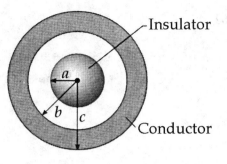

Figure P19.59

Solution

(a) Choose as the gaussian surface a concentric sphere of radius r. The electric field will be perpendicular to its surface, and will be uniform in strength over its surface.

The sphere of radius $r < a$ encloses charge
$$\rho\left(\tfrac{4}{3}\pi r^3\right)$$

so $\Phi = \dfrac{q}{\epsilon_0}$ becomes $E\left(4\pi r^2\right) = \dfrac{\rho\left(\tfrac{4}{3}\pi r^3\right)}{\epsilon_0}$ and $E = \dfrac{\rho\pi}{3\epsilon_0}$ ◊

For $a < r < b$, we have $E\left(4\pi r^2\right) = \dfrac{4}{3}\rho\dfrac{\pi a^3}{\epsilon_0} = \dfrac{Q}{\epsilon_0}$, and $E = \dfrac{\rho a^3}{3\epsilon_0 r^2} = \dfrac{Q}{4\pi\epsilon_0 r^2}$ ◊

For $b < r < c$, we must have $E = 0$ ◊

because any nonzero field would be moving charges in the metal. Free charges did move in the metal to deposit charge $-Q_b$ on its inner surface, at radius b, leaving charge $+Q_c$ on its outer surface, at radius c. Since the shell as a whole is neutral, $Q_c - Q_b = 0$.

For $r > c$, $\Phi = \dfrac{q}{\epsilon_0}$ becomes $E\left(4\pi r^2\right) = \dfrac{Q + Q_c - Q_b}{\epsilon_0}$

and $E = \dfrac{Q}{4\pi\epsilon_0 r^2}$ ◊

(b) For a gaussian surface of radius $b < r < c$, we have $0 = \dfrac{Q - Q_b}{\epsilon_0}$

so $Q_b = Q$ and the charge density on the inner surface is $\dfrac{-Q_b}{A} = \dfrac{-Q}{4\pi b^2}$ ◊

Then $Q_c - Q_b = 0$; the charge density on the outer surface is $+\dfrac{Q}{4\pi c^2}$ ◊

61. Repeat the calculations for Problem 60 when both sheets have **positive** uniform charge densities of value σ.

Solution The new, modified problem states:

"Two infinite, nonconducting sheets of charge are parallel to each other, as shown in Figure P19.60. The sheet on the left has a uniform surface charge density σ, and the one on the right has a uniform charge density σ. Calculate the electric field at points (a) to the left of, (b) in between, and (c) to the right of the two sheets."

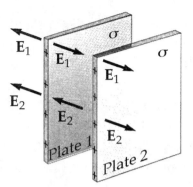

**Figure P19.60
(modified)**

For each sheet, the magnitude of the field at any point is

$$|\mathbf{E}| = \frac{\sigma}{2\epsilon_0}$$

(a) At a point to the left of the two parallel sheets

$$\mathbf{E} = E_1(-\mathbf{i}) + E_2(-\mathbf{i}) = 2E(-\mathbf{i})$$

$$\mathbf{E} = -\frac{\sigma}{\epsilon_0}\mathbf{i} \qquad \Diamond$$

(b) At a point between the two sheets

$$\mathbf{E} = E_1\mathbf{i} + E_2(-\mathbf{i}) = 0$$

$$\mathbf{E} = 0 \qquad \Diamond$$

(c) At a point to the right of the two parallel sheets

$$\mathbf{E} = E_1\mathbf{i} + E_2\mathbf{i} = 2E\mathbf{i}$$

$$\mathbf{E} = \frac{\sigma}{\epsilon_0}\mathbf{i} \qquad \Diamond$$

Chapter 20

Electric Potential and Capacitance

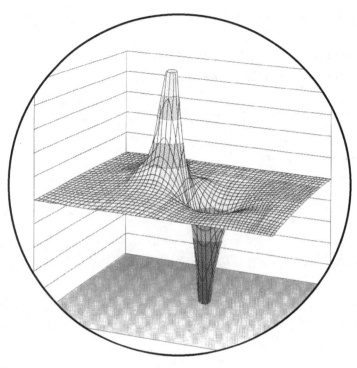

INTRODUCTION

The concept of potential energy was first introduced in Chapter 7 of the text, in connection with such conservative forces as the force of gravity and the elastic force of a spring. By using the law of energy conservation, we were often able to avoid working directly with forces when solving various mechanical problems. In this chapter we see that the energy concept is also of great value in the study of electricity. Since the electrostatic force given by Coulomb's law is conservative, electrostatic phenomena can be described conveniently in terms of an electrical potential energy. This idea enables us to define a scalar quantity called electric potential. Because the potential is a scalar function of position, it offers a simpler way of describing electrostatic phenomena than does the electric field.

This chapter is also concerned with the properties of capacitors, devices that store charge. Capacitors are commonly used in a variety of electrical circuits. For instance, they are used (1) to tune the frequency of radio receivers, (2) as filters in power supplies, (3) to eliminate sparking in automobile ignition systems, and (4) as energy-storing devices in electronic flash units.

A capacitor basically consists of two conductors separated by an insulator. We shall see that the capacitance of a given device depends on its geometry and on the material separating the charged conductors, called a dielectric.

NOTES FROM SELECTED CHAPTER SECTIONS

20.1 Potential Difference and Electric Potential

The electric potential at an arbitrary point is the work required per unit charge to bring a **positive test charge from infinity to that point.**

The potential difference between two points $V_B - V_A$ equals the work per unit charge that an **external agent** must perform in order to move a test charge, q_0, from point A to point B **without** a change in kinetic energy.

20.2 Potential Differences in a Uniform Electric Field

Electric field lines always point in the direction of decreasing electric potential. When a positive charge moves along the direction of the electric field, the electric potential energy of the charge-field system decreases. An equipotential surface is any surface consisting of a continuous distribution of points having the same electric potential. Equipotential surfaces are perpendicular to electric field lines and no work is required to move any charge between two points on an equipotential surface.

20.3 Electric Potential and Electric Potential Energy Due to Point Charges

Equipotential surfaces for a point charge are a family of spheres concentric with the charge.

20.4 Obtaining Electric Field From Electric Potential

If the electric potential (which is a scalar) is known as a function of coordinates (x, y, z), the components of the electric field (a vector quantity) can be obtained by taking the negative derivative of the potential with respect to the coordinates.

20.6 Electric Potential of a Charged Conductor

The surface of any charged conductor in equilibrium is an equipotential surface. Also **(since the electric field is zero inside the conductor)**, the potential is constant everywhere inside the conductor and equal to its value at the surface.

20.7   Capacitance

The capacitance of a capacitor depends on the physical characteristics of the device (size, shape, and separation of plates and the nature of the dielectric medium filling the space between the plates). Since the potential difference between the plates is proportional to the quantity of charge on each plate, the value of the capacitance is independent of the charge on the capacitor.

20.8   Combinations of Capacitors

When two or more unequal capacitors are connected in **series**, they carry the same charge, but the potential differences are not the same. Their capacitances add as reciprocals, and the equivalent capacitance of the combination is always **less** than the smallest individual capacitor.

When two or more capacitors are connected in **parallel**, the potential difference across each is the same. The charge on each capacitor is proportional to its capacitance; hence the capacitances add directly to give the equivalent capacitance of the parallel combination.

20.9   Energy Stored in a Charged Capacitor

The electrostatic potential energy stored in a charged capacitor equals the work done in the charging process – moving charges from one conductor at a lower potential to another conductor at a higher potential.

20.10   Capacitors with Dielectrics

A dielectric is a nonconducting material characterized by a dimensionless parameter – the dielectric constant, κ. In general, the use of a dielectric has the following effects:

- It increases the capacitance.

- It increases the maximum operating voltage.

- It provides mechanical support for the two conductors.

EQUATIONS AND CONCEPTS

The **potential difference** between two points A and B in an electric field, $\Delta V = V_B - V_A$ is defined as the ratio of the change in the potential energy of the charge-field system to the value of the test charge when the test particle is moved between two points. The potential difference can be found by integrating $\mathbf{E} \cdot d\mathbf{s}$ along **any path** from A to B.

$$\Delta V = \frac{\Delta U}{q_0} = -\int_A^B \mathbf{E} \cdot d\mathbf{s} \qquad (20.3)$$

If the field is **u n i f o r m**, the potential difference depends only on the displacement d in the direction parallel to $\mathbf{E}$. Electric field lines always point in the direction of decreasing electric potential.

$$\Delta V = -E \int_A^B ds = -E d \qquad (20.6)$$

When a charge moves between two points in an electric field the **change in the electric potential energy**, ΔU, of the charge-field system depends on the sign and magnitude of the charge as well as on the change in potential, ΔV. In the special case where the electric field is uniform, the change in potential energy is proportional to the distance the charge moves along a direction parallel to the electric field. When a positive charge moves along the direction of the electric field, the electric potential energy of the charge-field system decreases.

$$\Delta U = q_0 \Delta V = -q_0 E d \qquad (20.7)$$

The **electric potential** at a point in the vicinity of one or more point charges is calculated in a manner which assumes that the potential is zero at infinity.

$$V = k_e \sum_i \frac{q_i}{r_i} \qquad (20.12)$$

For **a pair of charges** separated by a distance r the **electric potential energy** represents the work required to assemble the charges from an infinite separation. Hence, the negative of the potential energy equals the minimum work required to separate them by an infinite distance. The electric potential energy associated with a system of two charged particles is positive if the two charges have the same sign, and negative if they are of opposite sign.

$$U = k_e \frac{q_1 q_2}{r_{12}} \qquad (20.13)$$

If there are more than two charged particles in the system, the **total electric potential energy** is found by calculating U for each pair of charges and summing the terms algebraically.

$$U = \frac{k_e}{2} \sum_{i=1}^{N} \sum_{j=1}^{N} \frac{q_i q_j}{r_{ij}}$$

$$\text{when } i \neq j$$

If the scalar electric potential function throughout a region of space is known, then the vector electric field can be calculated from the potential function. The **components of the electric field** in rectangular coordinates are given in terms of partial derivatives of the potential.

$$E_x = -\frac{\partial V}{\partial x}$$

$$E_y = -\frac{\partial V}{\partial y}$$

$$E_z = -\frac{\partial V}{\partial z}$$

If the charge density has spherical symmetry, where the charge density depends only on the radial distance, r, then the electric field is radial.

$$E_r = -\frac{dV}{dr}\hat{\mathbf{r}} \qquad (20.16)$$

For a continuous charge distribution, the **electric potential** (relative to zero at infinity) can be calculated by integrating the contribution due to a charge element dq over the line, surface, or volume which contains all the charge. Here, as in the case of a continuous charge distribution, it is convenient to represent dq in terms of the appropriate charge density.

$$V = k_e \int_{\substack{\text{All} \\ \text{Charge}}} \frac{dq}{r} \qquad (20.18)$$

The **capacitance** of a capacitor is defined as the ratio of the charge on either conductor (or plate) to the magnitude of the potential difference between the conductors. **Capacitance is always a positive quantity.**

$$C \equiv \frac{Q}{\Delta V} \qquad (20.19)$$

The **capacitance of an air-filled parallel-plate capacitor** is proportional to the area of the plates and inversely proportional to the separation of the plates.

$$C = \frac{\epsilon_0 A}{d} \qquad (20.21)$$

When the region between the plates of a capacitor is completely filled by a material of dielectric constant κ, the capacitance increases by the factor κ.

$$C = \kappa \frac{\epsilon_0 A}{d} \qquad (20.33)$$

The **equivalent capacitance** of a **parallel combination** of capacitors is larger than any individual capacitor in the group.

$$C_{eq} = C_1 + C_2 + C_3 + \dots \qquad (20.24)$$

The **equivalent capacitance** of a **series combination** of capacitors is smaller than the smallest capacitor in the group.

$$\frac{1}{C_{eq}} = \frac{1}{C_1} + \frac{1}{C_2} + \frac{1}{C_3} + \ldots \qquad (20.28)$$

The **electrostatic energy** stored in the electrostatic field of a charged capacitor equals the work done (by a battery or other source) in charging the capacitor from $q = 0$ to $q = Q$.

$$U = \frac{Q^2}{2C} = \tfrac{1}{2}Q\Delta V = \tfrac{1}{2}C(\Delta V)^2 \qquad (20.29)$$

The **energy density**, energy per unit volume, at any point in an electrostatic field is proportional to the square of the electric field intensity at that point.

$$u_E = \tfrac{1}{2}\epsilon_0 E^2 \qquad (20.31)$$

SUGGESTIONS, SKILLS, AND STRATEGIES

The vector expression giving the components of the electric field **E** over a region can be obtained from the scalar function which describes the electric potential, V, over the region by using Equation 20.15 and similar expressions for the y and z components:

$$E_x = -\frac{\partial V}{\partial x} \qquad\qquad E_y = -\frac{\partial V}{\partial y} \qquad\qquad E_z = -\frac{\partial V}{\partial z}$$

The derivatives in the above expressions are called **partial derivatives**. This means that when the derivative is taken with respect to any one coordinate, any other coordinates that appear in the expression for the potential function are treated as constants.

Since the electrostatic force is a conservative force, the work done by the electrostatic force in moving a charge q from an initial point A to a final point B depends only on the location of the two points and is independent of the path taken between A and B. When calculating potential differences using the equation

$$V_B - V_A = -\int_A^B \mathbf{E} \cdot d\mathbf{s} \tag{20.3}$$

any path between A and B may be chosen to evaluate the integral; therefore you should select a path for which the evaluation of the "line integral" in Equation 20.3 will be as convenient as possible. For example; where A, B, and C are constants, if $\mathbf{E}$ is in the form

$$\mathbf{E} = Ax\,\mathbf{i} + By\,\mathbf{j} + Cz\,\mathbf{k}$$

The potential is integrated as

$$\int_A^B \mathbf{E} \cdot d\mathbf{s} = \int_A^B (Ax\,dx + By\,dy + Cz\,dz)$$

and Equation 20.3 becomes

$$V_B - V_A = -\left(A\int_{x_A}^{x_B} x\,dx + B\int_{y_A}^{y_B} y\,dy + C\int_{z_A}^{z_B} z\,dz \right)$$

$$V_B - V_A = \frac{A}{2}\left(x_A{}^2 - x_B{}^2\right) + \frac{B}{2}\left(y_A{}^2 - y_B{}^2\right) + \frac{C}{2}\left(z_A{}^2 - z_B{}^2\right)$$

Problem-Solving Hints for Electric Potentials:

- When working problems involving electric potential, remember that potential is a **scalar quantity** (rather than a vector quantity like the electric field), so there are no components to worry about. Therefore, when using the superposition principle to evaluate the electric potential at a point due to a system of point charges, you simply take the algebraic sum of the potentials due to each charge. However, you must keep track of signs. The potential created by each positive charge ($V = k_e q / r$) is positive, while the potential for each negative charge is negative.

- Just as in mechanics, only **changes** in electric potential are significant, hence the point where you choose the potential to be zero is arbitrary. When dealing with point charges or a finite-sized charge distribution, we usually define $V = 0$ to be at a point infinitely far from the charges. However, if the charge distribution itself extends to infinity, some other nearby point must be selected as the reference point.

- The electric potential at some point P due to a continuous distribution of charge can be evaluated by dividing the charge distribution into infinitesimal elements of charge dq located at a distance r from the point P. You then treat this element as a point charge, so that the potential at P due to the element is $dV = k_e dq / r$. The total potential at P is obtained by integrating dV over the entire charge distribution. In performing the integration for most problems, it is necessary to express dq and r in terms of a single variable. In order to simplify the integration, it is important to give careful consideration of the geometry involved in the problem.

- Another method that can be used to obtain the potential due to a finite continuous charge distribution is to start with the definition of the potential difference given by Equation 20.3. If $\mathbf{E}$ is known or can be obtained easily (say from Gauss's law), then the line integral of $\mathbf{E} \cdot d\mathbf{s}$ can be evaluated. An example of this method is given in Ex. 20.6.

- Once you know the electric potential at a point, it is possible to obtain the electric field at that point by remembering that **the electric field is equal to the negative of the derivative of the potential with respect to some coordinate.** Example 20.5 illustrates how to use this procedure.

- Note that V is used to denote the value of the electric **potential at a point** (as in Eqs. 20.11 and 20.18) while ΔV is used to denote the **potential difference between two specified points** (as in Eqs. 2.3 and 20.8). If the electric potential has values V_A and V_B at points A and B, the electric potential difference in moving from point A to point B will be $\Delta V = V_B - V_A$.

In practice, a variety of phrases are used to describe the potential difference between two points, the most common being "voltage." A voltage **applied** to a device or **across** a device has the same meaning as the potential difference across the device. For example, if we say that the voltage across a certain capacitor is 12 volts, we mean that the potential difference between the capacitor's plates is 12 volts.

Problem-Solving Hints for Capacitance:

- When analyzing a series-parallel combination of capacitors to determine the equivalent capacitance, you should make a sequence of circuit diagrams, which show the successive steps in the simplification of the circuit. At each step combine and calculate the equivalent capacitance of those capacitors which are in either simple-series or simple-parallel combination with each other. Continue the sequence of simplifying steps until the final circuit contains a single capacitor, the value of which is the equivalent capacitance of the original circuit. At each step, you know two of the three quantities: Q, ΔV, and C. You will be able to determine the remaining quantity using the relation $Q = C\Delta V$.

- When calculating capacitance, be careful with your choice of units. To calculate capacitance in farads, make sure that distances are in meters and use the SI value of ϵ_0. When checking consistency of units, remember that the units for electric fields are newtons per coulomb (N/C) or the equivalent volts per meter (V/m).

- When two or more unequal capacitors are connected in series, they carry the same charge, but their potential differences are not the same. The capacitances add as reciprocals, and the equivalent capacitance of the combination is always less than the smallest individual capacitor.

- When two or more capacitors are connected in parallel, the potential differences across them are the same. The charge on each capacitor is proportional to its capacitance; hence, the capacitances add directly to give the equivalent capacitance of the parallel combination.

- A dielectric increases capacitance by the factor κ (the dielectric constant) because induced surface charges on the dielectric reduce the electric field inside the material from E to E / κ.

- Be careful about problems in which you may be connecting or disconnecting a battery to a capacitor. It is important to note whether modifications to the capacitor are being made while the capacitor is connected to the battery or after it is disconnected. If the capacitor remains connected to the battery, the voltage across the capacitor necessarily remains the same (equal to the battery voltage), and the charge is proportional to the capacitance, **however it may be modified** (say, by insertion of a dielectric). On the other hand, if you disconnect the capacitor from the battery before making any modifications to the capacitor, then its charge remains the same. In this case, as you vary the capacitance, the voltage across the plates changes in inverse proportion to capacitance, according to $\Delta V = Q / C$.

REVIEW CHECKLIST

▷ Understand that each point in the vicinity of a charge distribution can be characterized by a scalar quantity called the electric potential, V. The values of this potential function over the region (a scalar) are related to the values of the electrostatic field, **E**, over the region (a vector field).

▷ Calculate the electric potential difference between any two points in a uniform **electric field**, and the electric potential difference between any two points in the vicinity of a **group of point charges**.

▷ Calculate the electric **potential energy** associated with a group of point charges.

▷ Obtain an expression for the electric field (a **vector** quantity) over a region of space if the scalar electric potential function for the region is known.

▷ Calculate the work done by an external force in moving a charge q between any two points in an electric field when (a) an expression giving the field as a function of position is known, or when (b) the charge distribution (either point charges or a continuous distribution of charge) giving rise to the field is known.

▷ Determine the equivalent capacitance of a network of capacitors in series-parallel combination and calculate the final charge on each capacitor and the potential difference across each when a known potential is applied across the combination.

▷ Calculate the capacitance, potential difference, and stored energy of a capacitor which is partially or completely filled with a **dielectric**.

ANSWERS TO SELECTED CONCEPTUAL QUESTIONS

2. Give a physical explanation of the fact that the potential energy of a pair of like charges is positive whereas the potential energy of a pair of unlike charges is negative.

Answer   You may remember from the chapter on gravitational potential energy that potential energy of a system is defined to be positive when positive work must have been performed by an external agent to change the system from an initial configuration, to which we assign a zero value of potential energy, to a final configuration. For example the system of a flag and the Earth has positive potential energy when the flag is raised if we define zero potential energy as the configuration with the flag at the ground, since positive work must be done by an external force in order to raise it from the ground to the top of the pole.

When assembling like charges from an infinite separation, a configuration for which we have defined the potential energy as having a value of zero, it takes work to move them closer together to some distance r; therefore energy is being stored in the system of charges, and the potential energy is positive.

When assembling unlike charges from an infinite separation, the charges tend to accelerate toward each other, and thus energy is released as they approach a separation of distance r. Therefore, the potential energy of a pair of unlike charges is negative.

□   □   □   □

10.   If the potential difference across a capacitor is doubled, by what factor does the energy stored change?

Answer   Since $U = C \Delta V^2/2$, doubling ΔV will quadruple the stored energy.

□   □   □   □

12.   You are asked to design a capacitor with a small size and large capacitance. What factors would be important in your design?

Answer   You should use a dielectric filled capacitor whose dielectric constant is very large. Furthermore, you should make the dielectric as thin as possible, keeping in mind that dielectric breakdown must also be considered.

□   □   □   □

SOLUTIONS TO SELECTED END-OF-CHAPTER PROBLEMS

1. (a) Calculate the speed of a proton that is accelerated from rest through a potential difference of 120 V. (b) Calculate the speed of an electron that is accelerated through the same potential difference.

Solution

(a) Energy of the proton-field system is conserved as the proton moves from high to low potential. We imagine the proton moving between two points with potentials of 120 V and 0 V:

$$K_i + U_i + \Delta E_{mech} = K_f + U_f$$

(Review the energy version of the isolated system model from Chapters 6 and 7 to the full extent necessary for you.)

$$0 + q\Delta V + 0 = \tfrac{1}{2}mv_p{}^2 + 0$$

$$\left(1.60 \times 10^{-19}\ \text{C}\right)(120\ \text{V})\left(\frac{1\ \text{J}}{1\ \text{V} \cdot \text{C}}\right) = \tfrac{1}{2}\left(1.67 \times 10^{-27}\ \text{kg}\right)v_p{}^2$$

$$v_p = 1.52 \times 10^5\ \text{m/s} \qquad \Diamond$$

(b) The electron will gain speed in moving the other way, from $V_i = 0$ to $V_f = 120\ \text{V}$:

$$K_i + U_i + \Delta E_{mech} = K_f + U_f$$

$$0 + 0 + 0 = \tfrac{1}{2}mv_e{}^2 + q\Delta V$$

$$0 = \tfrac{1}{2}\left(9.11 \times 10^{-31}\ \text{kg}\right)v_e{}^2 + \left(-1.60 \times 10^{-19}\ \text{C}\right)(120\ \text{J/C})$$

$$v_e = 6.49 \times 10^6\ \text{m/s} \qquad \Diamond$$

This is less than one-tenth the speed of light; we do not need to use the relativistic kinetic energy formula.

7. An electron moving parallel to the x axis has an initial speed of 3.70×10^6 m/s at the origin. Its speed is reduced to 1.40×10^5 m/s at the point $x = 2.00$ cm. Calculate the potential difference between the origin and that point. Which point is at the higher potential?

Solution   Use the energy version of the isolated system model to equate the energy of the electron-field system when the electron is at

$$x = 0 \qquad\qquad \text{and} \qquad\qquad x = 2.00 \text{ cm.}$$

The unknown will be the difference in potential $V_f - V_i$.

Thus, $K_i + U_i + \Delta E_{mech} = K_f + U_f$

becomes $\frac{1}{2} m v_i^2 + q V_i + 0 = \frac{1}{2} m v_f^2 + q V_f$

or $\frac{1}{2} m \left(v_i^2 - v_f^2 \right) = q \left(V_f - V_i \right)$

so $V_f - V_i = \dfrac{m \left(v_i^2 - v_f^2 \right)}{2q}$

Noting that the electron's charge is negative, and evaluating the potential,

$$V_f - V_i = \frac{\left(9.11 \times 10^{-31} \text{ kg}\right)\left[\left(3.70 \times 10^6 \text{ m/s}\right)^2 - \left(1.40 \times 10^5 \text{ m/s}\right)^2 \right]}{2\left(-1.60 \times 10^{-19} \text{ C}\right)} = -38.9 \text{ V} \qquad \Diamond$$

The negative sign means that the 2.00-cm location is lower in potential than the origin. A positive charge would slow in free flight toward higher voltage, but the negative electron slows as it moves into lower potential. The 2.00-cm distance was unnecessary information for this problem. **If the field were uniform, we could find the x-component from $\Delta V = -E_x d$.**

13. The three charges in Figure P20.13 are at the vertices of an isosceles triangle. Calculate the electric potential at the midpoint of the base, taking $q = 7.00 \ \mu C$.

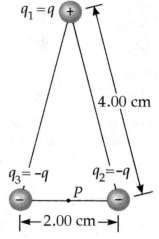

$q_1 = q$

4.00 cm

$q_3 = -q$ $q_2 = -q$

P

$\leftarrow 2.00 \text{ cm} \rightarrow$

Figure P20.13
(modified)

Solution Let $q_1 = q$ and $q_2 = q_3 = -q$. The charges are at distances of

$$r_1 = \sqrt{(0.0400 \text{ m})^2 - (0.0100 \text{ m})^2} = 3.87 \times 10^{-2} \text{ m}$$

and

$$r_2 = r_3 = 0.0100 \text{ m}$$

The potential at point P is

$$V_p = \frac{k_e q_1}{r_1} + \frac{k_e q_2}{r_2} + \frac{k_e q_3}{r_3} = k_e(q)\left(\frac{1}{r_1} + \frac{-1}{r_2} + \frac{-1}{r_3}\right)$$

so

$$V_p = (8.99 \times 10^9 \text{ N} \cdot \text{m}^2 / \text{C}^2)(7.00 \times 10^{-6} \text{ C})\left(\frac{1}{0.0387 \text{ m}} - \frac{1}{0.0100 \text{ m}} - \frac{1}{0.0100 \text{ m}}\right)$$

and

$$V_p = -11.0 \times 10^6 \text{ V} \qquad \Diamond$$

Related Calculation Calculate the electric field vector at the same point due to the three charges. The separate fields of the two negative charges are in opposite directions and add to zero:

$$\mathbf{E}_P = \frac{k_e q_1}{r_1^2} \hat{\mathbf{r}}_1 = \frac{(8.99 \times 10^9 \text{ N} \cdot \text{m}^2/\text{C}^2)(7 \times 10^{-6} \text{ C})}{(0.0400 \text{ m})^2 - (0.0100 \text{ m})^2} \text{ down}$$

$$\mathbf{E}_P = (42.0 \times 10^6 \text{ N/C})(-\mathbf{j})$$

17. Show that the amount of work required to assemble four identical point charges of magnitude Q at the corners of a square of side s is $5.41 k_e Q^2/s$.

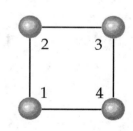

2   3

1   4

Solution   The work required equals the sum of the potential energies for all pairs of charges. No energy is involved in placing q_4 at a given position in empty space. When q_3 is brought from far away and placed close to q_4, the system potential energy can be expressed as $q_3 V_4$, where V_4 is the potential at the position of q_3 established by charge q_4. When q_2 is brought into the system, it interacts with two other charges, so we have two additional terms $q_2 V_3$ and $q_2 V_4$ in the total potential energy. Finally, when we bring the fourth charge q_1 into the system, it interacts with three other charges, giving us three more energy terms. Thus, the complete expression for the energy is:

$$U = q_1 V_2 + q_1 V_3 + q_1 V_4 + q_2 V_3 + q_2 V_4 + q_3 V_4$$

$$U = \frac{q_1 k_e q_2}{r_{12}} + \frac{q_1 k_e q_3}{r_{13}} + \frac{q_1 k_e q_4}{r_{14}} + \frac{q_2 k_e q_3}{r_{23}} + \frac{q_2 k_e q_4}{r_{24}} + \frac{q_3 k_e q_4}{r_{34}}$$

$$U = \frac{Q k_e Q}{s} + \frac{Q k_e Q}{s\sqrt{2}} + \frac{Q k_e Q}{s} + \frac{Q k_e Q}{s} + \frac{Q k_e Q}{s\sqrt{2}} + \frac{Q k_e Q}{s}$$

Evaluating, $$U = \frac{k_e Q^2}{s}\left(4 + \frac{2}{\sqrt{2}}\right) = 5.41 k_e \frac{Q^2}{s} \qquad \lozenge$$

---

25.   Over a certain region of space, the electric potential is $V = 5x - 3x^2 y + 2yz^2$. Find the expressions for x, y, and z components of the electric field over this region. What is the magnitude of the field at the point P, which has coordinates $(1, 0, -2)$ m?

Solution   First, we find the x, y, and z components of the field; then, we evaluate them at point P. (We assume that V is given in volts, as a function of distances in meters.)

$$E_x = -\frac{\partial V}{\partial x} = -5 + 6xy \qquad E_y = -\frac{\partial V}{\partial y} = 3x^2 - 2z^2 \qquad E_z = -\frac{\partial V}{\partial z} = -4yz \qquad \lozenge$$

At point P, $$E_x = -5 + 6(1.00\text{ m})(0\text{ m}) = -5.00\text{ N/C}$$

At point P, $$E_y = 3(1.00\text{ m})^2 - 2(-2.00\text{ m})^2 = -5.00\text{ N/C}$$

At point P, $$E_z = -4(0\text{ m})(-2.00\text{ m}) = 0\text{ N/C}$$

At P, the field's magnitude is $$E = \sqrt{(-5.00\text{ N/C})^2 + (-5.00\text{ N/C})^2 + 0^2} = 7.07\text{ N/C} \qquad \lozenge$$

---

27. A rod of length L (Fig. P20.27) lies along the x axis with its left end at the origin. It has a nonuniform charge density $\lambda = \alpha x$, where α is a positive constant. (a) What are the units of the α? (b) Calculate the electric potential at A.

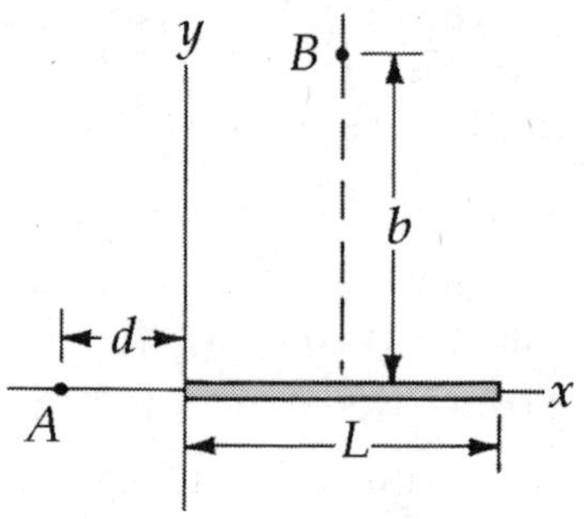

Solution

(a) As a linear charge density, λ has units of C/m. So $\alpha = \lambda/x$ must have units of C/m^2.   ◊

Figure P20.27

(b) Consider a small segment of the rod at location x and of length dx. The amount of charge on it is $\lambda\,dx = (\alpha x)\,dx$. Its distance from A is $d + x$, so its contribution to the electric potential at A is

$$dV = k_e \frac{dq}{r} = k_e \alpha x \frac{dx}{(d+x)}$$

We must integrate all these contributions for the whole rod, from $x = 0$ to $x = L$:

$$V = \int_{\text{all } q} dV = \int_0^L \frac{k_e \alpha x}{d+x}\,dx$$

To perform the integral, make a change of variables to $u = d + x$, $du = dx$, u (at $x = 0$) = d, and u (at $x = L$) = $d + L$:

$$V = \int_d^{d+L} \frac{k_e \alpha (u-d)}{u}\,du = k_e \alpha \int_d^{d+L} du - k_e \alpha d \int_d^{d+L} \left(\frac{1}{u}\right) du$$

[Keep track of symbols: the unknown is V. The values k_e, α, d, and L are known and constant. And x and u are variables, and will not appear in the answer.]

$$V = k_e \alpha u \Big|_d^{d+L} - k_e \alpha d \ln u \Big|_d^{d+L} = k_e \alpha (d+L-d) - k_e \alpha d (\ln(d+L) - \ln d)$$

$$V = k_e \alpha L - k_e \alpha d \ln\left(\frac{d+L}{d}\right)$$ ◊

We have the answer when the unknown is expressed in terms of the d, L, and α mentioned in the problem and the universal constant k_e.

31. A spherical conductor has a radius of 14.0 cm and a charge of 26.0 μC. Calculate the electric field and the electric potential at (a) $r = 10.0$ cm, (b) $r = 20.0$ cm, and (c) $r = 14.0$ cm from the center.

Solution

(a) Inside a conductor when charges are not moving, the electric field is zero and the potential is uniform, the same as on the surface.

$$\mathbf{E} = 0 \qquad \Diamond$$

$$V = \frac{k_e q}{R} = \frac{\left(8.99 \times 10^9 \text{ N} \cdot \text{m}^2 / \text{C}^2\right)\left(26.0 \times 10^{-6} \text{ C}\right)}{0.140 \text{ m}} = 1.67 \times 10^6 \text{ V} \qquad \Diamond$$

(b) The sphere behaves like a point charge at its center when you stand outside.

$$\mathbf{E} = \frac{k_e q}{r^2}\,\hat{\mathbf{r}} = \frac{\left(8.99 \times 10^9 \text{ N} \cdot \text{m}^2 / \text{C}^2\right)\left(26.0 \times 10^{-6} \text{ C}\right)}{(0.200 \text{ m})^2}\,\hat{\mathbf{r}} = \left(5.84 \times 10^6 \text{ N/C}\right)\hat{\mathbf{r}} \qquad \Diamond$$

$$V = \frac{k_e q}{r} = 1.17 \times 10^6 \text{ V} \qquad \Diamond$$

(c) $\mathbf{E} = \dfrac{k_e q}{r^2}\,\hat{\mathbf{r}} = \left(11.9 \times 10^6 \text{ N/C}\right)\hat{\mathbf{r}}$ $\qquad \Diamond$

$V = 1.67 \times 10^6$ V as in part (a) $\qquad \Diamond$

---

35. An air-filled capacitor consists of two parallel plates, each with an area of 7.60 cm^2, separated by a distance of 1.80 mm. If a 20.0-V potential difference is applied to these plates, calculate (a) the electric field between the plates, (b) the surface charge density, (c) the capacitance, and (d) the charge on each plate.

Solution

(a) The potential difference between two points in a uniform electric field is $\Delta V = Ed$, so:

$$E = \frac{\Delta V}{d} = \frac{20.0 \text{ V}}{1.80 \times 10^{-3} \text{ m}}$$

$$E = 1.11 \times 10^4 \text{ V/m} \qquad \qquad \Diamond$$

(b) The electric field between capacitor plates is $E = \dfrac{\sigma}{\epsilon_0}$

so $\sigma = \epsilon_0 E$:

$$\sigma = \left(8.85 \times 10^{-12} \text{ C}^2 / \text{N} \cdot \text{m}^2\right)\left(1.11 \times 10^4 \text{ V/m}\right)$$

$$\sigma = 9.83 \times 10^{-8} \text{ C/m}^2 = 98.3 \text{ nC/m}^2 \qquad \qquad \Diamond$$

(c) For a parallel-plate capacitor, $C = \dfrac{\epsilon_0 A}{d}$:

$$C = \frac{\left(8.85 \times 10^{-12} \text{ C}^2 / \text{N} \cdot \text{m}^2\right)\left(7.60 \times 10^{-4} \text{ m}^2\right)}{1.80 \times 10^{-3} \text{ m}}$$

$$C = 3.74 \times 10^{-12} \text{ F} = 3.74 \text{ pF} \qquad \qquad \Diamond$$

(d) The charge on each plate is $Q = C\Delta V$:

$$Q = \left(3.74 \times 10^{-12} \text{ F}\right)(20.0 \text{ V})$$

$$Q = 7.47 \times 10^{-11} \text{ C} = 74.7 \text{ pC} \qquad \qquad \Diamond$$

---

43. Four capacitors are connected as shown in Figure P20.43. (a) Find the equivalent capacitance between points a and b. (b) Calculate the charge on each capacitor if $\Delta V_{ab} = 15.0$ V.

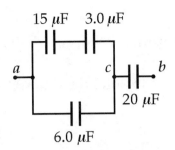

15 μF   3.0 μF

20 μF

6.0 μF

Figure P20.43

Solution

(a)   We simplify the circuit of Figure P20.43 in three steps as shown in figures (a), (b), and (c).

First, the 15.0 μF and 3.00 μF in series are equivalent to

$$\frac{1}{\left(\dfrac{1}{15.0\ \mu F} + \dfrac{1}{3.00\ \mu F}\right)} = 2.50\ \mu F$$

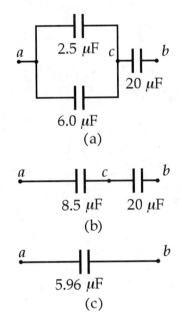

2.5 μF

20 μF

6.0 μF
(a)

Next, 2.50 μF combines in parallel with 6.00 μF, creating an equivalent capacitance of 8.50 μF.

8.5 μF   20 μF
(b)

At last, 8.50 μF and 20.0 μF are in series, equivalent to

$$\frac{1}{\left(\dfrac{1}{8.50\ \mu F} + \dfrac{1}{20.0\ \mu F}\right)} = 5.96\ \mu F \qquad \Diamond$$

5.96 μF
(c)

(b)   We find the charge on and the voltage across each capacitor by working backwards through solution figures (c)-(a), alternately applying $Q = C\Delta V$ and $\Delta V = Q/C$ to every capacitor, real or equivalent. For the 5.96 μF capacitor, we have

$$Q = C\Delta V = (5.96\ \mu F)(15.0\ V) = 89.5\ \mu C$$

Thus, if a is higher in potential than b, just 89.5 μC flows between the wires and the plates to charge the capacitors in each picture. In (b) we have, for the 8.5 μF capacitor,

$$\Delta V_{ac} = \frac{Q}{C} = \frac{89.5\ \mu C}{8.50\ \mu F} = 10.5\ V$$

and for the 20.0 μF in (b), (a), and the original circuit,

we have
$$Q_{20} = 89.5\ \mu C$$
◊

$$\Delta V_{cb} = \frac{Q}{C} = \frac{89.5\ \mu C}{20.0\ \mu F} = 4.47\ V$$

Next, (a) is equivalent to (b), so $\quad \Delta V_{cb} = 4.47\ V$

and $\qquad\qquad\qquad\qquad \Delta V_{ac} = 10.5\ V$

For the 2.50 μF, $\qquad\qquad \Delta V = 10.5\ V$

and $\qquad\qquad Q = C\Delta V = (2.50\ \mu F)(10.5\ V) = 26.3\ \mu C$

For the 6.00 μF, $\qquad\qquad \Delta V = 10.5\ V$

and $\qquad\qquad Q_6 = C\Delta V = (6.00\ \mu F)(10.5\ V) = 63.2\ \mu C$
◊

Now, 26.3 μC having flowed in the upper parallel branch in (a), back in the original circuit we have

$$Q_{15} = 26.3\ \mu C$$
◊

and $\qquad\qquad\qquad\qquad Q_3 = 26.3\ \mu C$
◊

Related Calculation   An exam problem might also ask for the voltage across each:

$$\Delta V_{15} = \frac{Q}{C} = \frac{26.3\ \mu C}{15.0\ \mu F} = 1.75\ V$$

and $\qquad\qquad \Delta V_3 = \frac{Q}{C} = \frac{26.3\ \mu C}{3.00\ \mu F} = 8.77\ V$
◊

45. Consider the circuit shown in Figure P20.45, where $C_1 = 6.00\ \mu F$, $C_2 = 3.00\ \mu F$, and $\Delta V = 20.0$ V. Capacitor C_1 is first charged by the closing of switch S_1. Switch S_1 is then opened, and the charged capacitor is connected to the uncharged capacitor by closing S_2. Calculate the initial charge acquired on C_1 and the final charge on each capacitor.

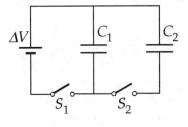

Figure P20.45

Solution

When S_1 is closed, the charge on C_1 will be

$$Q_1 = C_1 \Delta V_1 = (6.00\ \mu F)(20.0\ V) = 120\ \mu C \qquad \Diamond$$

When S_1 is opened and S_2 is closed, the total charge will remain constant and be shared by the two capacitors:

$$Q_1' = 120\ \mu C - Q_2'$$

The potential differences across the two capacitors will be equal.

$$\Delta V' = \frac{Q_1'}{C_1} = \frac{Q_2'}{C_2} \qquad \text{or} \qquad \frac{120\ \mu C - Q_2'}{6.00\ \mu F} = \frac{Q_2'}{3.00\ \mu F}$$

and $Q_2' = 40.0\ \mu C$ $\qquad \Diamond$

$$Q_1' = 120\ \mu C - 40.0\ \mu C = 80.0\ \mu C \qquad \Diamond$$

---

51. A parallel-plate capacitor has a charge Q and plates of area A. (a) What force acts on one plate to attract it toward the other plate? Because the electric field between the plates is $E = Q/A\epsilon_0$, you might think that the force is $F = QE = Q^2/A\epsilon_0$. This is wrong, because the field E includes contributions from both plates, and the field created by the positive plate cannot exert any other force on the positive plate. Show that the force exerted on each plate is actually $F = Q^2/2\epsilon_0 A$. (Hint: Let $C = \epsilon_0 A/x$ for an arbitrary plate separation x; then require that the work done in separating the two charged plates be $W = \int F dx$.) The force exerted on one charged plate by another is sometimes used in a machine shop to hold a workpiece stationary.

Solution

The electric field in the space between the plates is

$$E = \frac{\sigma}{\epsilon_0} = \frac{Q}{A\epsilon_0}$$

You might think that the force on one plate is

$$F = QE = \frac{Q^2}{A\epsilon_0}$$

but this is two times too large, because neither plate exerts a force on itself.

The force on one plate is exerted by the other, through its electric field:

$$E = \frac{\sigma}{2\epsilon_0} = \frac{Q}{2A\epsilon_0}$$

The force on each plate is:

$$F = (Q_{self})(E_{other}) = \frac{Q^2}{2A\epsilon_0}$$

To prove this, we follow the hint, and calculate that the work done in separating the plates, which equals the potential energy stored in the charged capacitor:

$$U = \frac{1}{2}\frac{Q^2}{C} = \int F\, dx$$

From the fundamental theorem of calculus,

$$dU = F\, dx$$

and

$$F = \frac{d}{dx}U = \frac{d}{dx}\left(\frac{Q^2}{2C}\right) = \frac{1}{2}\frac{d}{dx}\left(\frac{Q^2}{A\epsilon_0 / x}\right)$$

Solving,

$$F = \frac{1}{2}\frac{d}{dx}\left(\frac{Q^2 x}{A\epsilon_0}\right) = \frac{1}{2}\left(\frac{Q^2}{A\epsilon_0}\right) \qquad \lozenge$$

57. A parallel-plate capacitor is constructed using a dielectric material whose dielectric constant is 3.00 and whose dielectric strength is 2.00×10^8 V/m. The desired capacitance is 0.250 μF, and the capacitor must withstand a maximum potential difference of 4 000 V. Find the minimum area of the capacitor plates.

Solution $\qquad E_{max} = 2.00 \times 10^8 \text{ V/m} = \dfrac{\Delta V_{max}}{d} \qquad\qquad$ so $\qquad\qquad d = \dfrac{\Delta V_{max}}{E_{max}}$

For $\qquad C = \dfrac{\kappa \epsilon_0 A}{d} = 0.250 \times 10^{-6} \text{ F}$ and $\kappa = 3.00,$ $\qquad A = \dfrac{Cd}{\kappa \epsilon_0} = \dfrac{C \Delta V_{max}}{\kappa \epsilon_0 E_{max}}$

$$A = \frac{\left(0.250 \times 10^{-6} \text{ F}\right)\left(4000 \text{ V}\right)}{(3.00)\left(8.85 \times 10^{-12} \text{ F/m}\right)\left(2.00 \times 10^8 \text{ V/m}\right)} \qquad\qquad A = 0.188 \text{ m}^2 \qquad\qquad \lozenge$$

---

63. Calculate the work that must be done to charge a spherical shell of radius R to a total charge Q.

Solution

When the potential of the shell is V due to a charge q, the work required to add an additional increment of charge dq is

$$dW = Vdq \qquad\qquad \text{where} \qquad\qquad V = \frac{k_e q}{R}$$

$$dW = \left(\frac{k_e q}{R}\right) dq \qquad\qquad \text{and} \qquad\qquad W = \frac{k_e}{R}\int_0^Q q\,dq$$

Therefore, $\qquad\qquad\qquad\qquad\qquad\qquad\qquad\qquad\qquad W = \left(\frac{k_e}{R}\right)\left(\frac{Q^2}{2}\right) \qquad\qquad \lozenge$

Chapter 21

Current and Direct Current Circuits

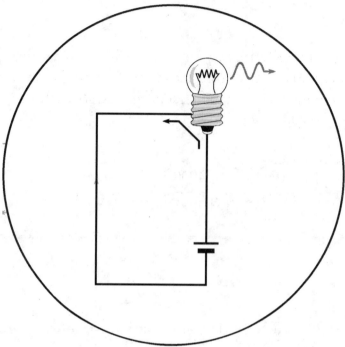

INTRODUCTION

Thus far our discussion of electrical phenomena has been confined to charges at rest, or electrostatics. We now consider situations involving electric charges in motion. The term **electric current,** or simply **current,** is used to describe the rate of flow of charge through some region of space.

In this chapter we first define current and current density. A microscopic description of current is given, and some of the factors that contribute to the resistance to the flow of charge in conductors are discussed. Mechanisms responsible for the electrical resistance of various materials depend on the composition of the material and on temperature. A classical model is used to describe electrical conduction in metals, and some of the limitations of this model are pointed out.

In this chapter we also analyze some simple circuits whose elements include batteries, resistors, and capacitors in various combinations. Such analysis is simplified by the use of two rules known as **Kirchhoff's rules,** which follow from the laws of conservation of energy and conservation of charge. Most of the circuits are assumed to be in **steady state,** which means that the currents are constant in magnitude and direction. We close the chapter with a discussion of circuits containing resistors and capacitors, in which current varies with time.

NOTES FROM SELECTED CHAPTER SECTIONS

21.1 Electric Current

The direction of conventional current is designated as the direction of motion of positive charge. In an ordinary metal conductor, the direction of current will be **opposite** the **direction of flow of electrons** (which are the charge carriers in this case).

21.2 Resistance and Ohm's Law

For ohmic materials, the ratio of the current density to the electric field (that gives rise to the current) is equal to a constant σ, the conductivity of the material. The reciprocal of the **conductivity** is called the **resistivity**, ρ. Each ohmic material has a characteristic resistivity, which depends only on the properties of the specific material and is a function of temperature.

21.4 A Structural Model for Electrical Conduction

In the classical model of electrical conduction in a metal, electrons are treated like molecules in a gas and, in the absence of an electric field, have a **zero average velocity**.

Under the influence of an electric field, the electrons move along a direction opposite the direction of the applied field with a **drift velocity**, which is proportional to the average time between collisions with atoms of the metal. The current is proportional to the magnitude of the drift velocity and to the number of free electrons per unit volume.

21.6 Sources of EMF

The emf of a source describes the work done by the source (e.g. battery or generator) as electrons move through the circuit. The emf of a battery is the potential difference between the terminals (terminal voltage) when there is no current in the battery. When there is current in the battery the terminal voltage is less than the emf by an amount, Ir where r is the **internal resistance** of the battery.

21.7   Resistors in Series and in Parallel

The **current** must be the same for each of a group of resistors connected in **series**.

The **potential difference** must be the same across each of a group of resistors in **parallel**.

21.8   Kirchhoff's Rules and Simple DC Circuits

- **Junction Rule**: The sum of the currents entering any junction must be equal to the sum of the currents leaving that junction. A junction is defined to be any point in the circuit where the current can split. To generate independent equations, the junction rule may be used a number of times which is **one fewer than the number of junction points in the circuit.**

- **Loop Rule**: The algebraic sum of the potential differences around any closed loop in a circuit must be zero. Each application of the loop rule should result in an **equation which contains a new circuit element (resistor or battery), or a new current.**

Overall, in order to solve a circuit problem using Kirchhoff's Rules, you must have as many independent equations as the total number of unknown quantities (currents, resistances, or emfs) in the circuit.

The first rule is a statement of **conservation of charge**; the second rule follows from the **conservation of energy**.

21.9   *RC* Circuits

Consider an uncharged capacitor in series with a resistor, a battery, and a switch. The charging process results in a net transfer of charge from one plate of the capacitor to the other moving along a path **through the resistor, battery, and switch**. The charges **do not move across the gap between the plates of the capacitor.**

The battery does work on the charges to increase their electrostatic potential energy as they move between the plates and the connecting wires.

EQUATIONS AND CONCEPTS

Under the action of an electric field, electric charges will move through gases, liquids, and solid conductors. **Electric current**, I, is defined as the rate at which charge moves through a cross section of the conductor.

$$I \equiv \lim_{t \to 0} \frac{\Delta Q}{\Delta t} = \frac{dQ}{dt} \qquad (21.2)$$

The direction of the current is in the direction of the flow of positive charges. The SI unit of current is the **ampere** (A).

$$1\,A = 1\,C/s \qquad (21.3)$$

The current in a conductor can be related to the number of mobile charge carriers per unit volume, n; the quantity of charge associated with each carrier, q; and the **drift velocity**, v_d, of the carriers.

$$I = \frac{\Delta Q}{\Delta t} = nqv_d A \qquad (21.4)$$

The **current density**, J, in a conductor is defined as the current per unit area.

$$J \equiv \frac{I}{A} = nqv_d \qquad (21.5)$$

For many practical applications, the definition of resistance relates the potential difference across a conductor and the current in the conductor to a composite of several physical characteristics of the conductor called the **resistance**, R.

$$R \equiv \frac{\Delta V}{I} \qquad (21.6)$$

The **resistance** of a given conductor of uniform cross section depends on the length, cross-sectional area, and a characteristic property of the material of which the conductor is made. The parameter, ρ, is the **resistivity** of the material of which the conductor is made. The resistivity is the inverse of the **conductivity** and has units of ohm-meters. The unit of resistance is the ohm (Ω).

$$R = \rho \frac{\ell}{A} \qquad (21.8)$$

$$1\,\Omega = 1\,\text{V}/\text{A}$$

The resistivity (and therefore the resistance) of a conductor varies with temperature in an approximately linear manner. In these expressions, α is the **temperature coefficient of resistivity** and T_0 is a stated reference temperature (usually 20°C).

$$\rho = \rho_0\left[1 + \alpha(T - T_0)\right] \qquad (21.10)$$

$$R = R_0\left[1 + \alpha(T - T_0)\right] \qquad (21.12)$$

The average time between collisions with atoms of a metal is an important parameter in the description of the classical model of electrical conduction in metals. This characteristic time is denoted by τ and can be related to the drift velocity (see Eq. 21.4) or the resistivity (see Eq. 21.11) associated with the conductor. In these equations, m_e and q represent the mass and charge of the electron, E is the magnitude of the applied electric field, and n is the number of free electrons per unit volume.

$$\mathbf{v}_d = \frac{-e\mathbf{E}}{m_e}\tau \qquad (21.15)$$

$$\rho = \frac{m_e}{ne^2\tau} \qquad (21.18)$$

Power will be supplied to a resistor or other current-carrying devices when a potential difference is maintained between the terminals of the circuit element. The quantities can be related in an equation called Joule's law and the SI unit of power is the watt (W). When the device is resistive, the power delivered can be expressed in alternative forms.

$$\mathcal{P} = I\Delta V \tag{21.20}$$

$$\mathcal{P} = I^2 R = \frac{(\Delta V)^2}{R} \tag{21.21}$$

When a battery is providing a current to an external circuit, the **terminal voltage**, ΔV, of the battery will be less than the emf due to **internal resistance** of the battery.

$$\Delta V = \mathcal{E} - Ir \tag{21.23}$$

The **current,** I, delivered by a battery in a simple **dc** circuit, depends on the value of the **emf** of the source, $\mathcal{E}$; the total **load resistance** in the circuit, R; and the **internal resistance** of the source, r.

$$\mathcal{E} = IR + Ir \tag{21.24}$$

The total or equivalent resistance of a series combination of resistors is equal to the sum of the resistances of the individual resistors.

$$R_{eq} = R_1 + R_2 + R_3 + \dots \tag{21.27}$$
(Series combination)

A group of resistors connected in parallel has an equivalent resistance, which is less than the smallest individual value of resistance in the group.

$$\frac{1}{R_{eq}} = \frac{1}{R_1} + \frac{1}{R_2} + \frac{1}{R_3} + \dots \tag{21.29}$$
(Parallel combination)

Resistors in series are connected so that pairs of adjacent resistors share only one common circuit point; there is a common current through each resistor in the group.

Resistors in parallel are connected so that each resistor in the parallel group has two circuit points in common with each of the other resistors; there is a common potential difference across each resistor in the group.

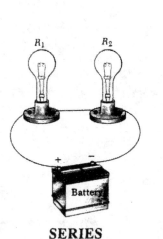
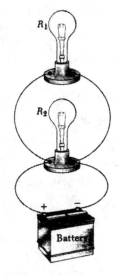

SERIES   **PARALLEL**

Many circuits, which contain several resistors, can be reduced to an equivalent single-loop circuit by successive step-by-step combinations of groups of resistors in series and parallel.

In the most general case, however, successive reduction is not possible and you must solve a true multiloop circuit by use of Kirchhoff's rules. Review the procedure suggested in the next section to apply Kirchhoff's rules.

When a potential difference is suddenly applied across an uncharged capacitor, the **current** in the circuit and the **charge** on the capacitor are functions of time. The **instantaneous values** of I and q depend on the capacitance and on the resistance in the circuit.

$$I(t) = \frac{\varepsilon}{R} e^{-t/RC} \qquad (21.34)$$

$$q(t) = Q\left[1 - e^{-t/RC}\right] \qquad (21.33)$$

When a battery is used to charge a capacitor in series with a resistor, a quantity, $\tau = RC$, called the time constant of the circuit, is used to describe the manner in which the charge on the capacitor varies with time. The charge on the capacitor increases from zero to 63% of its maximum value in a time interval equal to one time constant. Also, during one time constant, the charging current decreases from its initial maximum value of $I_0 = \mathcal{E}/R$ to 37% of I_0.

$$\tau = RC$$

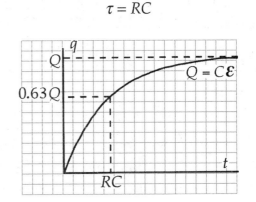

When a capacitor with an initial charge Q is discharged through a resistor, the **charge** and **current** decrease exponentially with time. The negative sign indicates the direction of the current, which is opposite the direction during the charging process.

$$q(t) = Qe^{-t/RC} \qquad (21.36)$$

$$I(t) = -I_0 e^{-t/RC} \qquad (21.37)$$

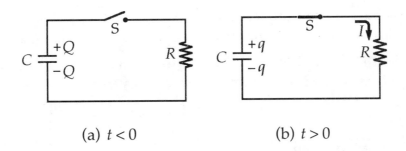

(a) $t < 0$ (b) $t > 0$

SUGGESTIONS, SKILLS, AND STRATEGIES

A Problem-Solving Strategy for Resistors:

- Be careful with your choice of units. To calculate the resistance of a device in ohms, make sure that distances are in meters and use the SI value of ρ.

- When two or more resistors are in **series** (connected end-to-end) they each carry the same current, but the potential differences across them are not equal (unless the resistors are equal). The resistors add directly to give the equivalent resistance of the series combination. The equivalent resistance will always be greater than any of the individual resistances.

- When two or more resistors are connected in **parallel**, the potential differences across them are the same. Since the current is inversely proportional to the resistance, the currents through them are not equal (unless the resistances are equal). The equivalent resistance of a parallel combination of resistors is found through reciprocal addition, and the equivalent resistor is always **less** than the smallest individual resistor.

- A complicated circuit consisting of resistors can often be reduced to a simple circuit containing only one resistor. To do so, repeatedly examine the circuit and replace any resistors that are in series or in parallel, using Eqs. 21.27 and 21.29. Sketch the new circuit after each set of changes has been made. Continue this process until a single equivalent resistance is found.

- If the current through, or the potential difference across, a resistor in the complicated circuit is to be identified, start with the final equivalent circuit found in the last step, and gradually work your way back through the circuits, using $\Delta V = IR$ at each step to find either the voltage drop or the current for each resistor in the group.

A Strategy for Using Kirchhoff's Rules:

- First, draw the circuit diagram and assign labels and symbols to all the known and unknown quantities. You must assign **directions** to the currents in each part of the circuit. Do not be alarmed if your assumption for a current's direction is wrong; the resulting value will be negative, but **its magnitude will be correct**. Although the assignment of current directions is arbitrary, you must stick with your guess throughout the problem as you apply Kirchhoff's rules. A resulting current with a negative value has a direction opposite that which you assumed.

- Apply the junction rule to any junction in the circuit. The junction rule may be applied as many times as a new current (one not used in a previous application) appears in the resulting equation. In general, the number of times the junction rule can be used is one fewer than the number of junction points in the circuit.

- Now apply Kirchhoff's loop rule to as many loops in the circuit as are needed to solve for the unknowns. Remember you must have as many equations as there are unknowns (I's, R's, and $\mathcal{E}$'s). In order to apply this rule, you must correctly identify the change in potential as you cross each element in traversing the closed loop. Watch out for signs!

The following figure illustrates convenient rules, which you may use to determine the increase or decrease in potential as the current in a loop crosses a resistor or seat of emf. Notice that the potential **decreases** (changes by $-IR$) when the resistor is traversed **in the direction of the current**. There is an **increase** in potential of $+IR$ if the direction of travel is **opposite** the direction of current. If a seat of emf is traversed **in** the direction of the emf (from – to + on the battery), the potential **increases** by $\mathcal{E}$. If the direction of travel is from + to –, the potential **decreases** by $\mathcal{E}$ (changes by $-\mathcal{E}$).

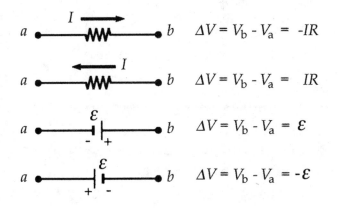

- Finally, you must solve the equations simultaneously for the unknown quantities. Be careful in your algebraic steps, and check your numerical answers for consistency.

As an illustration of the use of Kirchhoff's rules, consider a three-loop circuit, which has the **general form** shown in the following figure on the left. In this illustration, the actual circuit elements, Rs and $\mathcal{E}$s are not shown but assumed known. There are six possible different values of I in the circuit; therefore you will need six independent equations to solve for the six values of I. There are four junction points in the circuit (at points a, d, f, and h). The first rule applied at **any three** of these points will yield three equations. The circuit can be thought of as a group of three "blocks" or meshes as shown in the following figure on the right. Kirchhoff's second law, when applied to each of these loops ($abcda$, $ahfga$, and $defhd$), will yield three additional equations. You can then solve the total of six equations simultaneously for the six values of I_1, I_2, I_3, I_4, I_5, and I_6. You can, of course, expect that the sum of the changes in potential difference around **any other closed**

loop in the circuit will be zero (for example, $abcdefga$ or $ahfedcba$); however the equations found by applying Kirchhoff's second rule to these additional loops **will not be independent** of the six equations found previously.

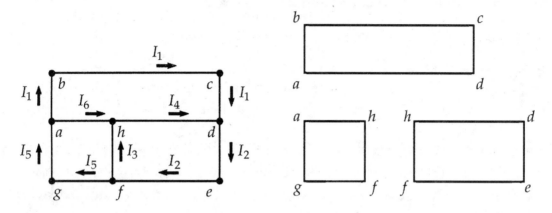

REVIEW CHECKLIST

▷ Define the term electric current in terms of rate of charge flow, and its corresponding unit of measure, the ampere. Calculate electron drift velocity, and the quantity of charge passing a point in a given time interval in a specified current-carrying conductor.

▷ Make calculations of the variation of resistance with temperature, which involves the concept of the temperature coefficient of resistivity.

▷ Determine the power delivered to a resistor.

▷ Calculate the equivalent resistance of a group of resistors in parallel, series, or series-parallel combination.

▷ Apply Kirchhoff's rules to solve multiloop circuits; that is, find the currents and the potential difference between any two points.

▷ Calculate the charging (discharging) current $I(t)$ and the accumulated (residual) charge $q(t)$ during charging (and discharging) of a capacitor in an RC circuit.

ANSWERS TO SELECTED CONCEPTUAL QUESTIONS

4. Two wires A and B of circular cross-section are made of the same metal and have equal lengths, but the resistance of wire A is three times greater than that of wire B. What is the ratio of their cross-sectional areas? How do the radii compare?

Answer Since $R = \dfrac{\rho \ell}{\text{Area}}$, the ratio of resistances is given by $\dfrac{R_A}{R_B} = \dfrac{\text{Area}_B}{\text{Area}_A}$

Hence, the ratio of their areas is three to one; that is, the area of wire B is three times that of wire A. From the ratio of the areas, we can calculate that the radius of wire B is $\sqrt{3}$ times the radius of wire A.

5. Use the atomic theory of matter to explain why the resistance of a material should increase as its temperature increases.

Answer As the temperature increases, the amplitude of atomic vibrations increases. This makes it more likely that the drifting electrons will be scattered by atomic vibrations, and makes it more difficult for charges to participate in organized motion inside the conductor.

8. If charges flow very slowly through a metal, why does it not require several hours for a light to come on when you throw a switch?

Answer Individual electrons move with a small average velocity through the conductor, but as soon as the voltage is applied, electrons all along the conductor start to move. Actually the current does not flow "immediately", but is limited by the speed of light.

13. Why is it possible for a bird to sit on a high-voltage wire without being electrocuted?

Answer

The resistance of the short segment of wire between the bird's feet is so small that the potential difference $\Delta V = IR$ between the feet is negligible. In order to be electrocuted, a potential difference is required. There is negligible potential difference between the bird's feet.

It could also be argued that a small amount of current does flow through the bird's feet, according to the rule of parallel resistors. However, since the resistance of the bird's feet is much higher than that of the wire, the amount of current that flows through the bird is still not enough to harm it.

21. In Figure Q21.21, describe what happens to the lightbulb after the switch is closed. Assume the capacitor has a large capacitance and is initially uncharged, and assume that the light illuminates when connected directly across the battery terminals.

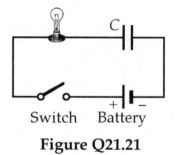

Figure Q21.21

Answer

The bulb will light up for an instant as the capacitor is being charged and there is a current in the circuit. As soon as the capacitor is fully charged, the current in the circuit will drop to zero, and the bulb will cease to glow.

SOLUTIONS TO SELECTED END-OF-CHAPTER PROBLEMS

5. Suppose that the current in a conductor decreases exponentially with time according to the equation

$$I(t) = I_0 e^{-t/\tau}$$

where I_0 is the initial current (at $t = 0$), and τ is a constant having dimensions of time. Consider a fixed observation point within the conductor. (a) How much charge passes this point between $t = 0$ and $t = \tau$? (b) How much charge passes this point between $t = 0$ and $t = 10\tau$? (c) How much charge passes this point between $t = 0$ and $t = \infty$?

Solution

From $I = \dfrac{dQ}{dt}$, we have $\qquad\qquad\qquad\qquad dQ = I\,dt$

From this, we derive a general integral $\qquad\qquad Q = \int dQ = \int I\,dt$

In all three cases, define an end-time, T. $\qquad\qquad Q = \int_0^T I_0 e^{-t/\tau}\,dt$

Integrating from time $t = 0$ to time $t = T$, $\qquad Q = \int_0^T (-I_0\tau) e^{-t/\tau}\left(-\dfrac{dt}{\tau}\right)$

Setting $Q = 0$ at $t = 0$, $\qquad\qquad\qquad\qquad Q = I_0\tau\left(1 - e^{-T/\tau}\right)$

(a) If $T = \tau$ $\qquad\qquad\qquad\qquad\qquad\qquad Q = -I_0\tau\left(e^{-1} - e^{0}\right)$

$\qquad\qquad\qquad\qquad\qquad\qquad\qquad\qquad Q = I_0\tau\left(e^{0} - e^{-1}\right) = 0.6321\, I_0\tau$ ◊

(b) If $T = 10\tau$ $\qquad\qquad\qquad\qquad\qquad Q = I_0\tau\left(1 - e^{-10}\right) = 0.99995\, I_0\tau$ ◊

(c) If $T = \infty$, $\qquad\qquad\qquad\qquad\qquad Q = I_0\tau\left(1 - e^{-\infty}\right) = I_0\tau$

9. A 0.900-V potential difference is maintained across a 1.50-m length of tungsten wire that has a cross-sectional area of 0.600 mm². What is the current in the wire?

Solution

From the definition of resistance, $I = \dfrac{\Delta V}{R}$ where $R = \dfrac{\rho L}{A}$

Therefore $I = \dfrac{\Delta V A}{\rho L} = \dfrac{(0.900 \text{ V})(6.00 \times 10^{-7} \text{ m}^2)}{(5.6 \times 10^{-8} \ \Omega \cdot \text{m})(1.50 \text{ m})} = 6.43 \text{ A}$ ◊

---

11. An aluminum wire with a diameter of 0.100 mm has a uniform electric field of 0.200 V/m imposed along its entire length. The temperature of the wire is 50.0 °C. Assume one free electron per atom. (a) Using the information given in Table 21.1, determine the resistivity. (b) What is the current density in the wire? (c) What is the total current in the wire? (d) What is the drift speed of the conduction electrons? (e) What potential difference must exist between the ends of a 2.00-m length of the wire if the stated electric field is to be produced?

Solution

The resistivity is found from $\rho = \rho_0[1 + \alpha(T - T_0)]$:

(a) $\rho = (2.82 \times 10^{-8} \ \Omega \cdot \text{m})[1 + (3.90 \times 10^{-3} \ °\text{C}^{-1})(30.0 \ °\text{C})] = 3.15 \times 10^{-8} \ \Omega \cdot \text{m}$ ◊

(b) $J = \sigma E = \dfrac{E}{\rho} = \left(\dfrac{0.200 \text{ V/m}}{3.15 \times 10^{-8} \ \Omega \cdot \text{m}}\right)(1 \ \Omega \cdot \text{A/V}) = 6.35 \times 10^6 \text{ A/m}^2$

(c) $J = \dfrac{I}{A} = \dfrac{I}{\pi r^2}$

$I = J \pi r^2 = (6.35 \times 10^6 \text{ A/m}^2)\left[\pi(5.00 \times 10^{-5} \text{ m})^2\right] = 49.9 \text{ mA}$ ◊

(d) The mass density gives the number-density of free electrons; we assume that each atom donates one free electron:

$$n = \left(2.70 \times 10^3 \text{ kg/m}^3\right)\left(\frac{1 \text{ mol}}{26.98 \text{ g}}\right)\left(\frac{10^3 \text{ g}}{\text{kg}}\right)\left(\frac{6.02 \times 10^{23} \text{ free e}^-}{1 \text{ mol}}\right) = 6.02 \times 10^{28} \text{ e}^- / \text{m}^3$$

Now $J = nqv_d$ gives:

$$v_d = \frac{J}{nq} = \frac{\left(6.35 \times 10^6 \text{ A/m}^2\right)}{\left(6.02 \times 10^{28} \text{ e}^- / \text{m}^3\right)\left(-1.60 \times 10^{-19} \text{ C/e}^-\right)} = -6.59 \times 10^{-4} \text{ m/s} \qquad \Diamond$$

The sign indicates that the electrons drift opposite to the field and current.

(e) $\Delta V = E\ell = (0.200 \text{ V/m})(2.00 \text{ m}) = 0.400 \text{ V}$ $\qquad \Diamond$

---

13. If the drift velocity of free electrons in a copper wire is 7.84×10^{-4} m / s, what is the electric field in the conductor?

Solution   Our structural model of electric conduction pictures a conduction electron as a particle in an electric field and as a particle under constant acceleration, during the time between its collisions with atoms. Nevertheless, by including the effects of collisions and by averaging over the zigzag motion of electrons, the structural model pictures electrons as a system of particles in equilibrium, with a well defined drift velocity. The density of conduction electrons in copper, from Example 21.1 of the text, is 8.48×10^{28} e$^-$ / m^3. The current density in the wire is

$$J = nqv_d = \left(8.48 \times 10^{28} \text{ e}^- / \text{m}^3\right)\left(1.60 \times 10^{-19} \text{ C/e}^-\right)\left(7.84 \times 10^{-4} \text{ m/s}\right) = 1.06 \times 10^7 \text{ A/m}^2$$

Applying Equation 21.17 to the current in Equation 21.5,

$$J = \sigma E = \frac{E}{\rho}$$

$$E = \rho J = \left(1.7 \times 10^{-8} \ \Omega \cdot \text{m}\right)\left(1.06 \times 10^7 \text{ A/m}^2\right) = 0.181 \text{ V/m} \qquad \Diamond$$

---

17. Suppose that a voltage surge produces 140 V for a moment. By what percentage does the output of a 120-V, 100-W light bulb increase? Assume its resistance does not change.

Solution   We find the resistance: $\mathcal{P}_1 = \Delta V_1 I_1$ so $I_1 = \dfrac{\mathcal{P}_1}{\Delta V_1} = \dfrac{100 \text{ W}}{120 \text{ V}} = 0.833 \text{ A}$

$$R = \frac{\Delta V_1}{I_1} = \frac{120 \text{ V}}{0.833 \text{ A}} = 144 \text{ }\Omega$$

Now the current is larger, $I_2 = \dfrac{\Delta V_2}{R} = \dfrac{140 \text{ V}}{144 \text{ }\Omega} = 0.972 \text{ A}$

and the power is much larger: $\mathcal{P}_2 = I_2 \Delta V_2 = (0.972 \text{ A})(140 \text{ V}) = 136 \text{ W}$

The percentage increase is $\dfrac{136 \text{ W} - 100 \text{ W}}{100 \text{ W}} = 0.361 = 36.1\%$ ◊

---

19. A certain toaster has a heating element made of Nichrome resistance wire. When the toaster is first connected to a 120-V source of potential difference (and the wire is at a temperature of 20.0 °C), the initial current is 1.80 A. The current begins to decrease as the resistive element warms up. When the toaster has reached its final operating temperature, the current has dropped to 1.53 A. (a) Find the power delivered to the toaster when it is at its operating temperature. (b) What is the final temperature of the heating element?

Solution   The resistor is a nonisolated system in steady state as it takes in energy by electric transmission and puts out energy by electromagnetic radiation and by energy transferred by heat to the adjacent air.

Most toasters are rated at about 1000 W (usually stamped on the bottom of the unit), so we might expect this one to have a similar power rating. The temperature of the heating element should be hot enough to toast bread but low enough that the nickel-chromium alloy element does not melt. (The melting point of nickel is 1455 °C, and chromium melts at 1907 °C.)

The power can be calculated directly by multiplying the current and the voltage. The temperature can be found from the linear conductivity equation for Nichrome, with $\alpha = 0.4 \times 10^{-3}$ °C^{-1} from Table 27.1.

(a) $\mathcal{P} = \Delta VI = (120 \text{ V})(1.53 \text{ A}) = 184 \text{ W}$ ◊

(b) The resistance at 20.0 °C is $R_0 = \dfrac{\Delta V}{I} = \dfrac{120 \text{ V}}{1.80 \text{ A}} = 66.7 \ \Omega$

At operating temperature, $R = \dfrac{120 \text{ V}}{1.53 \text{ A}} = 78.4 \ \Omega$

Neglecting thermal expansion, we have

$$R = \frac{\rho \ell}{A} = \frac{\rho_0 \left[1 + \alpha (T - T_0)\right] \ell}{A} = R_0 \left[1 + \alpha (T - T_0)\right]$$

$$T = T_0 + \frac{(R/R_0) - 1}{\alpha} = 20.0 \text{ °C} + \frac{(78.4 \ \Omega / 66.7 \ \Omega) - 1}{0.4 \times 10^{-3} \text{ °C}^{-1}} = 461 \text{ °C} \qquad ◊$$

Although this toaster appears to use significantly less power than most, the temperature seems high enough to toast a piece of bread in a reasonable amount of time. In fact, the temperature of a typical 1000-W toaster would only be slightly higher because Stefan's radiation law (Equation 17.35) tells us that the temperature might be about 700 °C. In either case, the operating temperature is well below the melting point of the heating element.

---

23. A battery has an emf of 15.0 V. The terminal voltage of the battery is 11.6 V when it is delivering 20.0 W of power to an external load resistor R. (a) What is the value of R? (b) What is the internal resistance of the battery?

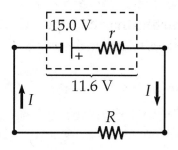

Solution   Combining the expression for power with the definition of resistance,

(a) $\mathcal{P} = \Delta VI$ and $\Delta V = IR$ give $R = \dfrac{\Delta V^2}{\mathcal{P}} = \dfrac{(11.6 \text{ V})^2}{20.0 \text{ W}} = 6.73 \ \Omega$ ◊

(b) $\mathcal{E} = IR + Ir$ so $r = \dfrac{\mathcal{E} - IR}{I}$

Because $I = \dfrac{\Delta V}{R}$, $r = \dfrac{(\mathcal{E} - \Delta V)R}{\Delta V}$ $r = \dfrac{(15.0 \text{ V} - 11.6 \text{ V})(6.73 \ \Omega)}{11.6 \text{ V}} = 1.97 \ \Omega$ ◊

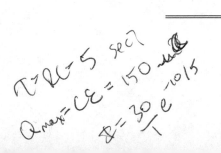

27. Consider the circuit shown in Figure P21.27. Find (a) the current in the 20.0-Ω resistor and (b) the potential difference between points a and b.

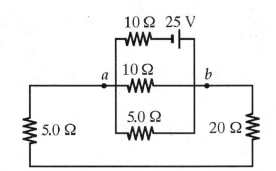

Figure P21.27

Solution If we turn the given diagram on its side, we find that it is the same as figure (a). The 20.0-Ω and 5.00-Ω resistors are in series, so the first reduction is as shown in (b). In addition, since the 10.0-Ω, 5.00-Ω, and 25.0-Ω resistors are then in parallel, we can solve for their equivalent resistance as:

$$R_{eq} = \frac{1}{(1/10.0\ \Omega) + (1/5.00\ \Omega) + (1/25.0\ \Omega)} = 2.94\ \Omega$$

This is shown in figure (c), which in turn reduces to the circuit shown in (d).

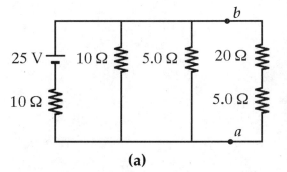

(a)

Next, we work backwards through the diagrams, applying $I = \Delta V / R$ and $\Delta V = IR$. The 12.94-Ω resistor is connected across 25.0-V, so the current through the voltage source in every diagram is

$$I = \frac{\Delta V}{R} = \frac{25.0\ \text{V}}{12.94\ \Omega} = 1.93\ \text{A}$$

In figure (c), this 1.93 A goes through the 2.94-Ω equivalent resistor to give a voltage drop of:

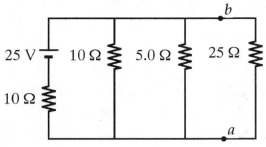

(b)

$$\Delta V = IR = (1.93\ \text{A})(2.94\ \Omega) = 5.68\ \text{V}$$

From figure (b), we see that this voltage drop is the same across ΔV_{ab}, the 10-Ω resistor, and the 5.00-Ω resistor.

(b) Therefore, $\Delta V_{ab} = 5.68\ \text{V}$ ◊

Since the current through the 20-Ω resistor is also the current through the 25.0-Ω line ab,

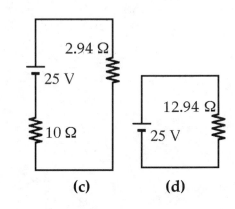

(c)   **(d)**

(a) $I = \dfrac{\Delta V_{ab}}{R_{ab}} = \dfrac{5.68\ \text{V}}{25.0\ \Omega} = 0.227\ \text{A}$ ◊

137

$I_1 + I_2 = I_3$

$12V + (.01\,\Omega)I_1 - (1.00\,\Omega)I_2 - 10V = 0$ $I_2 = +2 + .01I_1$

$10V + (1\Omega)I_2 + (.06\,\Omega)I_3 = 0$ (I_1I_2)

$10 + (2+.01I_1) + (.06)(I_1 + 2 + .01I_1) = 0$

$12 + .01I_1 + (.06)(.99I_1 + 2) = 0$

$8 + .01I_1 + .059I_1 - .12 = 0$ $I_1 = 158$

$3.0\,\Omega$ $I_2 = -9$

$I_3 = 1$

$\#17$

$.4057$

33. Determine the current in each branch of the circuit shown in Figure P21.33.

Solution First, we arbitrarily define the initial current directions and names, as shown in the second figure to the right.

The current rule then says that $I_3 = I_1 + I_2$ (1)

By the voltage rule, clockwise around the left-hand loop,

$+I_1(8.00\ \Omega) - I_2(5.00\ \Omega) - I_2(1.00\ \Omega) - 4.00\ V = 0$ (2)

and clockwise around the right-hand loop,

$4.00\ V + I_2(1.00\ \Omega + 5.00\ \Omega) + I_3(3.00\ \Omega + 1.00\ \Omega) - 12.0\ V = 0$ (3)

Solving by substitution rather than by determinants has the advantage that (just as when a cat has kittens) the answers, after the first, come out much more easily. Thus we substitute $(I_1 + I_2)$ for I_3, and reduce our three equations to:

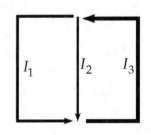

Figure P21.33

$$\begin{cases} (8.00\ \Omega)I_1 - (6.00\ \Omega)I_2 - 4.00\ V = 0 \\ 4.00\ V + (6.00\ \Omega)I_2 + (4.00\ \Omega)(I_1 + I_2) - 12.0\ V = 0 \end{cases} \text{ or } \begin{cases} I_2 = \dfrac{(8.00\ \Omega)I_1 - 4.00\ V}{6.00\ \Omega} \\ -8.0\ V + (4.0\ \Omega)I_1 + (10.0\ \Omega)I_2 = 0 \end{cases}$$

Solving the top equation for I_2, and then substituting I_2 into the equation below it,

$$-8.00\ V + (4.00\ \Omega)I_1 + \frac{10.0}{6.00}\left((8.00\ \Omega)I_1 - 4.00\ V\right) = 0$$

$$(17.3\ \Omega)I_1 - 14.7\ V = 0$$

and $I_1 = 0.846$ A down the 8 Ω resistor ◊

Thus, $I_2 = \dfrac{(8.00\ \Omega)(0.846\ A) - 4.00\ V}{6.00\ \Omega} = 0.462$ A down in middle branch ◊

$I_3 = 0.846\ A + 0.462\ A = 1.31$ A up in the right-hand branch ◊

$12 + .01I_1 + .0617I_1 + .122 = 0$

$I_1 = 171!$

$I_2 = .283)$

39. Consider a series RC circuit (see Fig. 21.26) for which $R = 1.00$ MΩ, $C = 5.00\ \mu$F, and $\mathcal{E} = 30.0$ V. Find (a) the time constant of the circuit and (b) the maximum charge on the capacitor after the switch is closed. (c) If the switch is closed at $t = 0$, find the current in the resistor 10.0 s later.

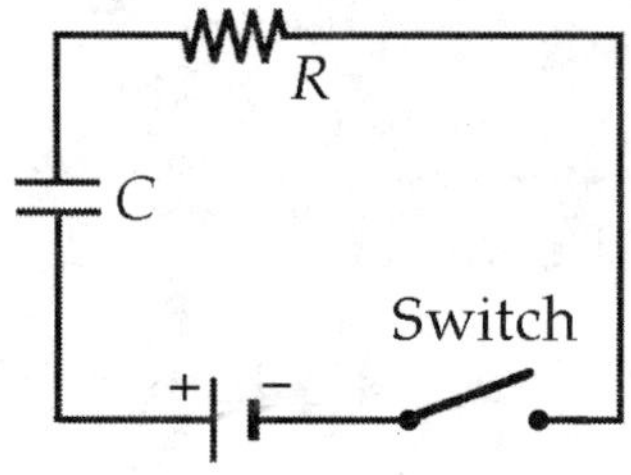

Figure 21.26

Solution

(a) $\tau = RC = \left(1.00 \times 10^6\ \Omega\right)\left(5.00 \times 10^{-6}\ \text{F}\right) = 5.00\ \Omega \cdot \text{F} = 5.00$ s   ◊

(b)   After a long time, the capacitor is "charged to thirty volts," separating charges

$$Q = C\Delta V = \left(5.00 \times 10^{-6}\right)(30.0\ \text{V}) = 150\ \mu\text{C}$$ ◊

(c) $I = I_0 e^{-t/\tau}$.   where $I_0 = \dfrac{\mathcal{E}}{R}$ and $\tau = RC = 5.00$ s

$$I = \frac{\mathcal{E}}{R}e^{-t/\tau} = \left(\frac{30.0\ \text{V}}{1.00 \times 10^6\ \Omega}\right)e^{-10.0\ \text{s}/5.00\ \text{s}}$$

$I = 4.06 \times 10^{-6}\ \text{A} = 4.06\ \mu\text{A}$ ◊

---

41. The circuit in Figure P21.41 has been connected for a long time. (a) What is the voltage across the capacitor? (b) If the battery is disconnected, how long does it take the capacitor to discharge to one-tenth of its initial voltage?

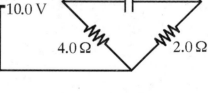

Figure P21.41

Solution

(a)   After a long time the capacitor branch will carry negligible current. The current flow is as shown in Figure (a).

To find the voltage at point a, we first find the current, using the voltage rule:

$$10.0 \text{ V} - (1.00 \ \Omega)I_2 - (4.00 \ \Omega)I_2 = 0$$

$$I_2 = 2.00 \text{ A}$$

$$V_a - V_0 = (4.00 \ \Omega)I_2 = 8.00 \text{ V}$$

Similarly, $\quad 10.0 \text{ V} - (8.00 \ \Omega)I_3 - (2.00 \ \Omega)I_3 = 0$

$$I_3 = 1.00 \text{ A}$$

At point b, $\quad V_b - V_0 = (2.00 \ \Omega)I_3 = 2.00 \text{ V}$

Thus, the voltage across the capacitor is

$$V_a - V_b = 8.00 \text{ V} - 2.00 \text{ V} = 6.00 \text{ V} \qquad \lozenge$$

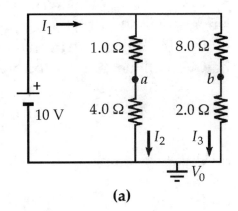

(a)

(b) We suppose the battery is pulled out leaving an open circuit. We are left with Figure (b), which can be reduced to equivalent circuits (c) and (d).

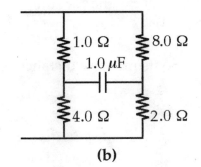

(b)

From (d), we can see that the capacitor discharges through a 3.60 Ω equivalent resistance. According to $q = Qe^{-t/RC}$,

we calculate that $\qquad qC = QCe^{-t/RC}$

and $\qquad \Delta V = \Delta V_i e^{-t/RC}$

Solving, $\qquad \dfrac{1}{10}\Delta V_i = \Delta V_i e^{-t/(3.60 \ \Omega)(1.00 \ \mu F)}$

$$e^{-t/3.60 \ \mu s} = 0.100$$

$$(-t/3.60 \ \mu s) = \ln(0.100) = -2.30$$

$$\dfrac{t}{3.60 \ \mu s} = 2.30$$

$$t = (2.30)(3.60 \ \mu s) = 8.29 \ \mu s$$

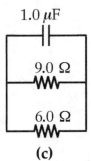

(c)

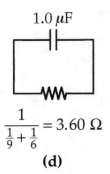

$$\dfrac{1}{\frac{1}{9} + \frac{1}{6}} = 3.60 \ \Omega$$

(d) $\qquad\qquad \lozenge$

51. An electric car is designed to run off a bank of 12.0-V batteries with a total energy storage of 2.00×10^7 J. (a) If the electric motor draws 8.00 kW, what is the current in the motor? (b) If the electric motor draws 8.00 kW as car moves at a steady speed of 20.0 m/s, how far will the car travel before it is "out of juice"?

Solution

(a) Since $\mathcal{P} = \Delta V I$

$$I = \frac{\mathcal{P}}{\Delta V} = \frac{8.00 \times 10^3 \text{ W}}{12.0 \text{ V}} = 667 \text{ A} \qquad \Diamond$$

(b) The time the car runs is

$$\Delta t = \frac{U}{\mathcal{P}} = \left(\frac{2.00 \times 10^7 \text{ J}}{8.00 \times 10^3 \text{ W}} \right) \left(\frac{1 \text{ W} \cdot \text{s}}{\text{J}} \right) = 2.50 \times 10^3 \text{ s}$$

So its moves a distance of

$$\Delta x = v \Delta t = (20.0 \text{ m/s})(2.50 \times 10^3 \text{ s}) = 50.0 \text{ km} \qquad \Diamond$$

Chapter 22

Magnetic Forces and Magnetic Fields

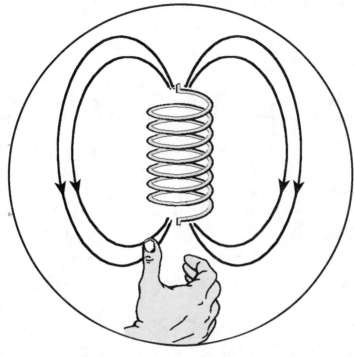

INTRODUCTION

Many historians of science believe that the compass, which uses a magnetic needle, was used in China as early as the 13th century BC, its invention being of Arab or Indian origin. The early Greeks knew about magnetism as early as 800 BC. They discovered that certain stones, now called magnetite (Fe_3O_4), attract pieces of iron. In 1269 Pierre de Maricourt mapped out the directions taken by a needle when it was placed at various points on the surface of a spherical natural magnet. He found that the directions formed lines that encircled the sphere and passed through two points diametrically opposite to each other, which he called the **poles** of the magnet. Subsequent experiments showed that every magnet, regardless of its shape, has two poles, called **north** and **south poles**, which exhibit forces on each other in a manner analogous to electric charges. That is, like poles repel each other and unlike poles attract each other.

This chapter deals with the origin of the magnetic field, namely, moving charges or electric currents. We show how to use the Biot-Savart law to calculate the magnetic field produced at a point by a current element. Using this formalism and the superposition principle, we then calculate the total magnetic field due to a current distribution. Next, we show how to determine the force between two current-carrying conductors, a calculation that leads to the definition of the ampere. We also introduce Ampère's law, which is very useful for calculating the magnetic field of highly symmetric configurations carrying steady currents. We apply Ampère's law to determine the magnetic field for a long current-carrying wire, a solenoid, and a toroid.

NOTES FROM SELECTED CHAPTER SECTIONS

22.2　The Magnetic Field

Particles with charge q, moving with speed v in a magnetic field $\mathbf{B}$, experience a magnetic force $\mathbf{F}_B$. Properties of the magnetic force are:

- The magnetic force is proportional to the charge q and speed v of the particle.
- The magnitude and direction of the magnetic force depend on the angle between the velocity vector of the particle and the magnetic field vector.
- When a charged particle moves in a direction **parallel** to the magnetic field vector, the magnetic force $\mathbf{F}_B$ on the charge is **zero**.
- The magnetic force acts in a direction perpendicular to both $\mathbf{v}$ and $\mathbf{B}$; that is, $\mathbf{F}_B$ is perpendicular to the plane formed by $\mathbf{v}$ and $\mathbf{B}$.
- The magnetic force on a positive charge is in the direction opposite to the force on a negative charge moving in the same direction.
- If the velocity vector makes an angle θ with the magnetic field, the magnitude of the magnetic force is proportional to $\sin \theta$.

There are several important differences between electric and magnetic forces:

- The electric force is always in the direction of the electric field, whereas the magnetic force is perpendicular to the magnetic field.
- The electric force acts on a charged particle independent of the particle's velocity, whereas the magnetic force acts on a charged particle only when the particle is in motion.
- The electric force does work in displacing a charged particle, whereas the magnetic force associated with a steady magnetic field does **no** work when a particle is displaced.

22.7   The Biot-Savart Law

The **Biot-Savart law** says that if a wire carries a steady current I, the magnetic field $d\mathbf{B}$ at a point P associated with an element $d\mathbf{s}$ has the following properties:

- The vector $d\mathbf{B}$ is perpendicular both to $d\mathbf{s}$ (which is in the direction of the current) and to the unit vector $\hat{\mathbf{r}}$ (directed from the current element to the point P.)

- The magnitude of $d\mathbf{B}$ is inversely proportional to r^2, where r is the distance from the element to the point P.

- The magnitude of $d\mathbf{B}$ is proportional to the current and to the length of the element, $d\mathbf{s}$.

- The magnitude of $d\mathbf{B}$ is proportional to $\sin\theta$, where θ is the angle between the vectors $d\mathbf{s}$ and $\hat{\mathbf{r}}$.

22.8   The Magnetic Force Between Two Parallel Conductors

Parallel conductors carrying currents in the **same direction attract** each other, whereas parallel conductors carrying currents in **opposite directions repel** each other.

The force between two parallel wires each carrying a current is used to define the ampere as follows:

If two long, parallel wires 1 m apart in a vacuum carry the same current and the force per unit length on each wire is 2×10^{-7} N/m, then the current is defined to be 1 A.

22.9   Ampère's Law

The direction of the magnetic field due to a current in a conductor is given by the right-hand rule:

If the wire is grasped in the right hand with the thumb in the direction of the current, the fingers will wrap (or curl) in the direction of **B**.

Ampère's law is valid only for **steady** currents and is useful only in those cases where the current configuration has a **high degree** of **symmetry**.

EQUATIONS AND CONCEPTS

The magnetic field (or magnetic induction) at some point in space is defined in terms of the **magnetic force** exerted on a moving positive electric charge at that point. The SI unit of the magnetic field is the tesla (T) or weber per square meter (Wb/m²).

$$\mathbf{F}_B = q\mathbf{v} \times \mathbf{B} \tag{22.1}$$

$$1\,\text{T} = 1\,\frac{\text{N} \cdot \text{s}}{\text{C} \cdot \text{m}}$$

The magnetic force will be of maximum magnitude when the charge moves along a direction perpendicular to the direction of the magnetic field. In general, the velocity vector may be directed along some direction other than 90.0° relative to the magnetic field. In this case, the magnetic force on the moving charge is less than its maximum value.

$$F_B = |q|vB\sin\theta \tag{22.2}$$

Equation 22.2 can be written in a form, which serves to define the magnitude of the magnetic field.

$$B \equiv \frac{F_B}{|q|v\sin\theta}$$

In order to determine the direction of the magnetic force, apply the right-hand rule:

> Hold your right hand with your fingers first pointing in the direction of **v** and then curling in the direction of **B**, as shown in the figure. The magnetic force on a positive charge points in the direction of your thumb. If the charge is negative, then the direction of the force is reversed.

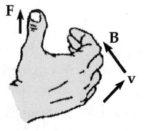

If a **straight** wire carrying a current is placed in an external magnetic field, a **magnetic force** will be exerted on the wire. The magnetic force on a wire of arbitrary shape is found by integrating over the length of the wire. In these equations the direction of $\boldsymbol{\ell}$ and $d\mathbf{s}$ is that of the current.

$$\mathbf{F}_B = I\boldsymbol{\ell} \times \mathbf{B} \qquad (22.10)$$

$$\mathbf{F}_B = I\int_a^b d\mathbf{s} \times \mathbf{B} \qquad (22.12)$$

The magnitude of a magnetic force on a current-carrying conductor in a magnetic field depends on the angle between the direction of the conductor and the direction of the field.

$$F_B = BI\ell \sin\theta$$

The magnetic force will be maximum when the conductor is directed perpendicular to the magnetic field.

$$F_{B,\,max} = BI\ell$$

When a closed conducting loop carrying a current is placed in an external magnetic field, there is a **net torque** exerted on the loop. In Equation 22.14, the area vector $\mathbf{A}$ is directed perpendicular to the area of the loop with a sense given by the right-hand rule. The magnitude of $\mathbf{A}$ is numerically equal to the area of the loop.

$$\boldsymbol{\tau} = I\mathbf{A} \times \mathbf{B} \qquad (22.14)$$

The magnitude of the torque will depend on the angle between the direction of the magnetic field and the direction of the normal (or perpendicular) to the plane of the loop.

$$\tau = IAB\sin\theta$$

The magnitude of the torque will be maximum when the magnetic field is parallel to the plane of the loop.

$$\tau_{max} = IAB \qquad (22.13)$$

The torque on a current loop can also be expressed in terms of μ, the magnetic dipole moment of the loop.

$$\tau = \mu \times B \qquad (22.16)$$

where $\mu = IA$ $\qquad (22.15)$

The direction of rotation of the loop is such that the magnetic moment of the loop turns in a direction so as to be more lined up with the magnetic field.

The **Biot-Savart law** gives the magnetic field at a point in space due to a small length element of conductor ds which carries a current I and is at a distance r away from the point.

$$d\mathbf{B} = \frac{\mu_0}{4\pi}\frac{I\,d\mathbf{s} \times \hat{\mathbf{r}}}{r^2} \qquad (22.20)$$

The permeability of free space is a constant.

$$\mu_0 = 4\pi \times 10^{-7}\ \text{T} \cdot \text{m}/\text{A}$$

The **total magnetic field** is found by integrating the Biot-Savart law expression over the entire length of the conductor.

$$\mathbf{B} = \frac{\mu_0 I}{4\pi} \int\limits_{\substack{\text{All}\\\text{Current}}} \frac{d\mathbf{s} \times \hat{\mathbf{r}}}{r^2}$$

The magnitude of the magnetic field due to several important geometric arrangements of a current-carrying conductor can be calculated by use of the Biot-Savart law:

B at a distance r from a **long straight conductor**, carrying a current I.

$$B = \frac{\mu_0 I}{2\pi r}$$

(22.21)

B at the center of an **arc of radius** R **which subtends an angle** θ **(in radians) at the center of the arc.**

$$B = \frac{\mu_0 I}{4\pi R}\theta$$

B_x on the axis of a **circular loop** of radius R and at a **distance** x **from the plane** of the loop.

$$B_x = \frac{\mu_0 I R^2}{2\left(x^2 + R^2\right)^{3/2}}$$

(22.23)

Substituting $x = 0$ gives the magnetic field at the center of the loop.

$$B = \frac{\mu_0 I}{2R}$$

(22.24)

The magnitude of the **magnetic force per unit length** between very long parallel conductors depends on the distance a between the conductors and the magnitudes of the two currents.

$$\frac{F}{\ell} = \frac{\mu_0}{2\pi}\frac{I_1 I_2}{a}$$

(22.27)

If the parallel currents I_1 and I_2 are in the same direction, the force between conductors will be one of attraction. Parallel conductors carrying currents in opposite directions will repel each other. In any case, the magnitude of the forces on the two conductors will be **equal.**

Ampère's law represents a relationship between the integral of the tangential component of the magnetic field around any closed path and the total current that the path encloses. Results for two example current configurations are:

$$\oint \mathbf{B} \cdot d\mathbf{s} = \mu_0 I \qquad (22.29)$$

B **inside a toroid** having N turns and at a distance r from the center of the toroid.

$$B = \frac{\mu_0 N I}{2\pi r} \qquad (22.31)$$

B near the center of a **solenoid** of n turns per unit length.

$$B = \mu_0 n I \qquad (22.32)$$

The direction of the magnetic field due to a current in a long wire is determined by using the right-hand rule for **B**:

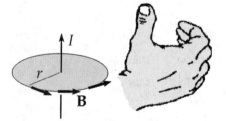

Hold the conductor in the right hand with the thumb pointing in the direction of the conventional current. The fingers will then wrap around the wire in the direction of the magnetic field lines. The magnetic field is tangent to the circular field lines at every point in the region around the conductor.

The direction of the magnetic field at the center of a current loop is perpendicular to the plane of the loop and directed in the sense given by the right-hand rule for **B**.

For the current shown, **B** is directed out of the paper.

Within a solenoid, the magnetic field is parallel to the axis of the solenoid and pointing in a sense determined by applying the right-hand rule for **B** to one of the coils.

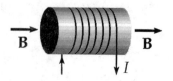

SUGGESTIONS, SKILLS, AND STRATEGIES

To remember the symbols for vectors that point away from you and towards you, think of a three-dimensional archery arrow as it turns to point first away from you (shown on the left), and then towards you, (shown on the right). The point on the arrow is represented by a dot; the feathers are represented by an 'x'.

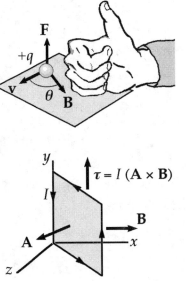

Equation 22.1, $\mathbf{F}_B = q\mathbf{v} \times \mathbf{B}$, serves as the definition of the magnetic field vector $\mathbf{B}$. The direction of the magnetic force $\mathbf{F}$ is determined by the right-hand rule for the cross product as illustrated in the figure to the right. (This assumes that the charge q is a positive charge.)

The right-hand rule and the vector cross product are also used to determine the direction of the torque and the direction of the resulting rotation for a closed current loop in a magnetic field. When the four fingers on your right hand circle the current loop in the direction of the current, the thumb will point in the direction of the area vector $\mathbf{A}$. Applying this rule to the situation shown in the figure to the right, the vector $\mathbf{A}$ is directed out of the plane of the rectangular loop facing you as shown.

The resulting torque is parallel to the direction of $\mathbf{A} \times \mathbf{B}$ and is along the positive y axis as shown in the figure. This is consistent with the general rule for determining the direction of the cross product of two vectors. If the loop is considered to be hinged along the edge joining the y axis, then the rotation will be such that the angle θ between $\mathbf{A}$ and $\mathbf{B}$ decreases. The loop rotates until its area vector $\mathbf{A}$ is parallel to the magnetic field vector $\mathbf{B}$. This is shown in the figure as a counterclockwise rotation about the y axis as seen from above.

It is important to remember that the **Biot-Savart law**, given by Equation 22.20,

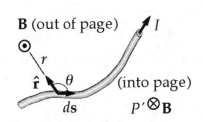

$$dB = \frac{\mu_0}{4\pi} \frac{I\, d\mathbf{s} \times \hat{\mathbf{r}}}{r^2}$$

is a **vector** expression. The unit vector $\hat{\mathbf{r}}$ is directed from the element of conductor $d\mathbf{s}$ to the point P where the magnetic field is to be calculated, and r is the distance from $d\mathbf{s}$ to point P. For the arbitrary current element shown in the figure at right, the direction of $\mathbf{B}$ at point P, as determined by the right-hand rule for the cross product, is directed **out of the plane**; while the magnetic field at point P' due to the current in the element $d\mathbf{s}$ is directed **into the plane**. In order to find the **total** magnetic field at any point due to a conductor, you must sum up the contributions from all current elements making up the conductor. This means that the total **B** field is expressed as an integral over the entire length of the conductor:

$$\mathbf{B} = \frac{\mu_0 I}{4\pi} \int_{length} \frac{d\mathbf{s} \times \hat{\mathbf{r}}}{r^2}$$

REVIEW CHECKLIST

▷ Use the defining equation for a magnetic field **B** to determine the magnitude and direction of the magnetic force exerted on an electric charge moving in a region where there is a magnetic field. You should understand clearly the important differences between the forces exerted on electric charges by electric fields and those forces exerted on moving electric charges by magnetic fields.

▷ Calculate the magnitude and direction of the magnetic force on a current-carrying conductor when placed in an external magnetic field.

▷ Determine the magnitude and direction of the torque exerted on a closed current loop in an external magnetic field. You should understand how to designate the direction of the area vector corresponding to a given current loop, and to incorporate the magnetic moment of the loop into the calculation of the torque on the loop.

▷ Use the Biot-Savart law to calculate the magnetic field at a specified point in the vicinity of a current element, and by integration find the total magnetic field due to a number of important geometric arrangements. Your use of the Biot-Savart law must include a clear understanding of the **direction** of the magnetic field contribution relative to the direction of the current element, which produces it and the direction of the vector, which locates the point at which the field is to be calculated.

▷ Use Ampère's law to calculate the magnetic field due to steady current configurations, which have a sufficiently high degree of symmetry such as a long straight conductor, a long solenoid, and a toroidal coil.

ANSWERS TO SELECTED CONCEPTUAL QUESTIONS

1. Two charged particles are projected into a region where a magnetic field is perpendicular to their velocities. If the charges are deflected in opposite directions, what can you say about them?

Answer We know the magnetic field is constant, and the velocity vectors are the same, but one force is the negative of the other. From $\mathbf{F}_B = q(\mathbf{v} \times \mathbf{B})$, we can conclude that the only thing that could cause the force to be of opposite sign is if the charges were of opposite sign.

3. Is it possible to orient a current loop in a uniform magnetic field so that the loop does not tend to rotate? Explain.

Answer Yes. If the magnetic field is perpendicular to the plane of the loop, the forces on opposite sides will be equal and opposite, but will produce no net torque.

13. Explain why two parallel wires carrying currents in opposite directions repel each other.

Answer The figure at the right will help you understand this result. The magnetic field due to wire 2 at the position of wire 1 is directed out of the paper. Hence, the magnetic force on wire 1, given by $I_1\mathbf{L}_1 \times \mathbf{B}_2$, must be directed to the left since $\mathbf{L}_1 \times \mathbf{B}_2$ is directed to the left. Likewise, you can show that the magnetic force on wire 2 due to the field of wire 1 is directed towards the right.

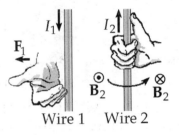

Wire 1   Wire 2

14. A hollow copper tube carries a current along its length. Why is $\mathbf{B} = 0$ inside the tube? Is $\mathbf{B}$ nonzero outside the tube?

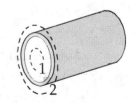

Answer Let us apply Ampere's law to the closed path labeled 1 in this figure. Since there is no current through this path, and because of the symmetry of the configuration, we see that the magnetic field inside the tube must be zero. On the other hand, the net current through the path labeled 2 is I, the current carried by the conductor. Therefore, the field outside the tube is nonzero.

SOLUTIONS TO SELECTED END-OF-CHAPTER PROBLEMS

1. Determine the initial direction of the deflection of charged particles as they enter the magnetic fields as shown in Figure P22.1.

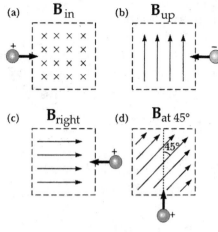

Solution

(a) By solution figure (a), $\mathbf{v} \times \mathbf{B}$ is $(\text{right}) \times (\text{away}) = \text{up}$ ◊

(b) By solution figure (b), $\mathbf{v} \times \mathbf{B}$ is $(\text{left}) \times (\text{up}) = \text{away}$. Since the charge is negative, $q\mathbf{v} \times \mathbf{B}$ is toward you   ◊

(c) $\mathbf{v} \times \mathbf{B}$ is zero since the angle between $\mathbf{v}$ and $\mathbf{B}$ is $180°$ and $\sin 180° = 0$. There is no deflection.   ◊

Figure P22.1

(d) $\mathbf{v} \times \mathbf{B}$ is $(\text{up}) \times (\text{up and right})$, or away from you   ◊

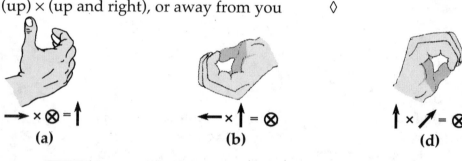

7. A cosmic-ray proton in interstellar space has an energy of 10.0 MeV and executes a circular orbit having a radius equal to that of Mercury's orbit around the Sun (5.80×10^{10} m). What is the magnetic field in that region of space?

Solution   Think of the proton as having accelerated through a potential difference $\Delta V = 10^7$ V. We use the energy version of the isolated system model, applied to the particle and the electric field that made it speed up from rest.

The proton's kinetic energy is

$$E = \frac{1}{2}mv^2 = e\Delta V$$

so its speed is

$$v = \sqrt{\frac{2e\Delta V}{m}}$$

Now we use the particle in a magnetic field model and the particle in uniform circular motion model.

$\Sigma F = ma$ becomes

$$\frac{mv^2}{R} = evB\sin 90°$$

So

$$B = \frac{mv}{eR} = \frac{m}{eR}\sqrt{\frac{2e\,\Delta V}{m}} = \frac{1}{R}\sqrt{\frac{2m\Delta V}{e}}$$

and

$$B = \frac{1}{5.80\times 10^{10}\ \text{m}}\sqrt{\frac{2\left(1.6727\times 10^{-27}\ \text{kg}\right)\left(10^7\ \text{V}\right)}{1.60\times 10^{-19}\ \text{C}}} = 7.88\times 10^{-12}\ \text{T} \quad \Diamond$$

11. The picture tube in a television uses magnetic deflection coils rather than electric deflection plates. Suppose an electron beam is accelerated through a 50.0-kV potential difference and then through a region of uniform magnetic field 1.00 cm wide. The screen is located 10.0 cm from the center of the coils and is 50.0 cm wide. When the field is turned off, the electron beam hits the center of the screen. What field magnitude is necessary to deflect the beam to the side of the screen? Ignore relativistic corrections.

Solution   The beam is deflected by the angle

$$\theta = \tan^{-1}\left(\frac{25.0\ \text{cm}}{10.0\ \text{cm}}\right) = 68.2°$$

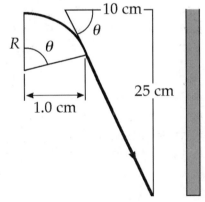

The two angles θ shown are equal because their sides are perpendicular, right side to right side and left side to left side. The radius of curvature of the electrons in the field is

$$R = \frac{1.00\ \text{cm}}{\sin 68.2°} = 1.077\ \text{cm}$$

Now $\frac{1}{2}mv^2 = q\Delta V$ so $v = \sqrt{\frac{2q\Delta V}{m}} = 1.33\times 10^8\ \text{m/s}$

α $r = \dfrac{mv}{qB_0}$ $v = \dfrac{\mathcal{E}}{B}$

$\Sigma \mathbf{F} = m\mathbf{a}$ becomes

$$\frac{mv^2}{R} = |q|vB\sin 90°$$

$$B = \frac{mv}{|q|R} = \frac{\left(9.11 \times 10^{-31} \text{ kg}\right)\left(1.33 \times 10^8 \text{ m/s}\right)}{\left(1.60 \times 10^{-19} \text{ C}\right)\left(1.077 \times 10^{-2} \text{ m}\right)} = 70.1 \text{ mT} \qquad \lozenge$$

15.   A nonuniform magnetic field exerts a net force on a magnetic dipole. A strong magnet is placed under a horizontal conducting ring of radius r that carries current I, as shown in Figure P22.15. If the magnetic field **B** makes an angle θ with the vertical at the ring's location, what are the magnitude and direction of the resultant force on the ring?

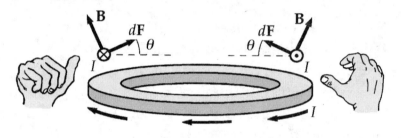

Figure P22.15

Solution   The magnetic force on each bit of ring is inward and upward, at an angle θ above the radial line, according to:

$$|d\mathbf{F}| = I|d\mathbf{s} \times \mathbf{B}| = I\,ds\,B$$

The radially inward components tend to squeeze the ring, but cancel out as forces. The upward components $I\,ds\,B\sin\theta$ all add to

$$\mathbf{F} = I(2\pi r)B\sin\theta \text{ up} \qquad \lozenge$$

The magnetic moment of the ring is down. This problem is a model for the force on a dipole in a nonuniform magnetic field, or for the force that one magnet exerts on another magnet.

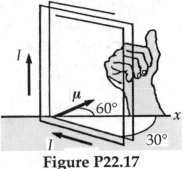

(handwritten notes at top)
$$27)\ A$$
$$B_A = B_1 \cos 45 +$$
$$B_2 \cos 45 + B_3$$
$$= \frac{\mu_0 I}{2\pi a}\left[\frac{2}{\sqrt{2}}\cos 45 + \frac{1}{3}\right]$$
$$B)\ B_3 = \frac{\mu_0 I}{2\pi (2a)}$$

17. A rectangular coil consists of $N = 100$ closely wrapped turns and has dimensions $a = 0.400$ m and $b = 0.300$ m. The coil is hinged along the y axis, and its plane makes an angle $\theta = 30.0°$ with the x axis (Fig. P22.17). What is the magnitude of the torque exerted on the coil by a uniform magnetic field $B = 0.800$ T directed along the x axis when the current is $I = 1.20$ A in the direction shown? What is the expected direction of rotation of the coil?

Figure P22.17
(modified)

Solution The magnetic moment of the coil is $\mu = NIA$, perpendicular to its plane and making a 60° angle with the x axis as shown to the right. The torque on the dipole is then

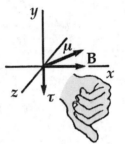

$$\boldsymbol{\tau} = \boldsymbol{\mu} \times \mathbf{B} = NBAI\sin\theta \ \text{down}$$

having a magnitude of $\qquad \tau = NBAI\sin\theta$

$$\tau = (100)(0.800\ \text{T})(0.400 \times 0.300\ \text{m}^2)(1.20\ \text{A})\sin 60.0° = 9.98\ \text{N}\cdot\text{m} \quad \lozenge$$

Note that θ is the angle between the magnetic moment and the **B** field. We model the coil as a rigid body under a net torque; the coil will rotate so as to align the magnetic moment with the **B** field. Looking down along the y axis, we will see the coil rotate in the clockwise direction. $\qquad\qquad\qquad \lozenge$

---

25. Determine the magnetic field at a point P located a distance x from the corner of an infinitely long wire bent at a right angle, as in Figure P22.25. The wire carries a steady current I.

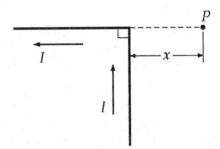

Solution The vertical section of wire constitutes one half of an infinitely long straight wire at distance x from P, so it creates a field equal to

Figure P22.25

$$B = \frac{1}{2}\left(\frac{\mu_0 I}{2\pi x}\right)$$

(handwritten at bottom left)
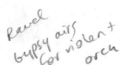
$$27)\ \frac{\mu_0 I}{4\pi}\left(\frac{1}{r^2}\right)$$

(handwritten at bottom right) Ravel
gypsy airs
for violent
orch

Hold your right hand with extended thumb in the direction of the current; the field is away from you, into the paper. For each bit of the horizontal section of wire $d\mathbf{s}$ is to the left and $\hat{\mathbf{r}}$ is to the right, so $d\mathbf{s} \times \hat{\mathbf{r}} = 0$. The horizontal current produces zero field at P. Thus,

$$\mathbf{B} = \frac{\mu_0 I}{4\pi x} \text{ into the paper}$$

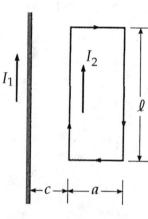

◊

---

31. In Figure P22.31, the current in the long, straight wire is $I_1 = 5.00$ A and the wire lies in the plane of the rectangular loop, which carries 10.0 A. The dimensions are $c = 0.100$ m, $a = 0.150$ m, and $\ell = 0.450$ m. Find the magnitude and direction of the net force exerted on the loop by the magnetic field created by the wire.

Solution

By symmetry, the forces exerted on the segments of length a are equal and opposite and cancel.

The magnetic field in the plane of I_2 to the right of I_1 is directed away from you into the plane.

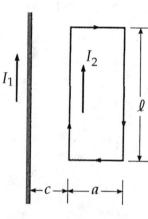

Figure P22.31

By the right-hand rule, $\mathbf{F}_B = I\boldsymbol{\ell} \times \mathbf{B}$ is directed toward the **left** for the near side of the loop and directed toward the **right** for the side at $c + a$. Thus,

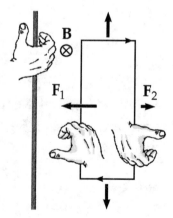

$$\mathbf{F} = \mathbf{F}_1 + \mathbf{F}_2 = \frac{\mu_0 I_1 I_2 \ell}{2\pi}\left(\frac{1}{c+a} - \frac{1}{c}\right)\mathbf{i} = \frac{\mu_0 I_1 I_2 \ell}{2\pi}\left(\frac{-a}{c(c+a)}\right)\mathbf{i}$$

$$\mathbf{F} = \frac{\left(4\pi \times 10^{-7}\ \text{N/A}^2\right)(5.00\ \text{A})(10.0\ \text{A})(0.450\ \text{m})}{2\pi}\left(\frac{-0.150\ \text{m}}{(0.100\ \text{m})(0.250\ \text{m})}\right)\mathbf{i}$$

$$\mathbf{F} = \left(-2.70 \times 10^{-5}\,\mathbf{i}\right)\ \text{N} \qquad \text{or} \qquad \mathbf{F} = 2.70 \times 10^{-5}\ \text{N toward the left} \qquad ◊$$

---

33. Four long, parallel conductors carry equal currents of $I = 5.00$ A. Figure P22.33 is an end view of the conductors. The current direction is into the page at points A and B (indicated by the crosses) and out of the page at C and D (indicated by the dots). Calculate the magnitude and direction of the magnetic field at point P, located at the center of the square with an edge length of 0.200 m.

A(×)---------(•)*C*

P

0.200 m

B(×)---------(•)*D*

0.200 m

Figure P22.33

Solution

Each wire is distant from P by $(0.200 \text{ m})\cos 45.0° = 0.141$ m.

Each wire produces a field at P of equal magnitude:

$$B = \frac{\mu_0 I}{2\pi a} = \frac{(2.00 \times 10^{-7} \text{ T} \cdot \text{m/A})(5.00 \text{ A})}{0.141 \text{ m}} = 7.07 \ \mu\text{T}$$

Carrying currents away from you, the left-hand wires produce fields at P of 7.07 μT, in the following directions:

A: to the bottom and left, at 225°
B: to the bottom and right, at 315°;

Carrying currents toward you, the wires to the right also produce fields at P of 7.07 μT, in the following directions:

C: downward and to the right, at 315°
D: downward and to the left, at 225°.

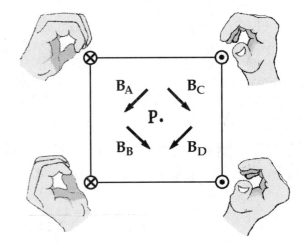

The total field is then

$$4(7.07 \ \mu\text{T})\sin 45.0° = 20.0 \ \mu\text{T} \text{ toward the bottom of the page.}$$

37. A packed bundle of 100 long, straight, insulated wires forms a cylinder of radius $R = 0.500$ cm. (a) If each wire carries 2.00 A, what are the magnitude and direction of the magnetic force per unit length acting on a wire located 0.200 cm from the center of the bundle? (b) Would a wire on the outer edge of the bundle experience a force greater or less than the value calculated in part (a)?

Solution

The force **on one** wire is exerted **by the other ninety-nine,** through the magnetic field they create.

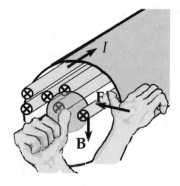

(a) According to Equation 22.30 in Example 22.7, the magnetic field at $r = 0.200$ cm from the center of the cable is:

$$B = \frac{\mu_0 I_0 r}{2\pi R^2} = \frac{\left(4\pi \times 10^{-7}\ \text{T} \cdot \text{m/A}\right)(99)(2.00\ \text{A})\left(0.200 \times 10^{-2}\ \text{m}\right)}{2\pi \left(0.500 \times 10^{-2}\ \text{m}\right)^2}$$

$$B = 3.17 \times 10^{-3}\ \text{T}$$

This field points tangent to a circle of radius 0.2 mm. It will exert a force $\mathbf{F}_B = I\boldsymbol{\ell} \times \mathbf{B}$ toward the center of the bundle, on the hundredth wire:

$$\frac{F_B}{\ell} = I B \sin\theta = (2.00\ \text{A})\left(3.17 \times 10^{-3}\ \text{T}\right)\sin 90° = 6.34\ \text{mN/m} \text{ toward the center} \qquad \lozenge$$

(b) As is shown in Figure 22.31 of the text, the field is strongest at the outer surface of the cable, so the force on one strand is greater here, by a factor of $5/2$. $\qquad \lozenge$

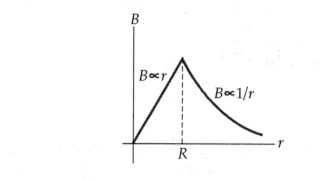

41. What current is required in the windings of a long solenoid that has 1000 turns uniformly distributed over a length of 0.400 m, to produce at the center of the solenoid a magnetic field of magnitude 1.00×10^{-4} T?

Solution

$$B = \mu_0 \frac{N}{\ell} I \qquad \text{so} \qquad I = \frac{B\ell}{\mu_0 N} = \frac{\left(1.00 \times 10^{-4} \text{ T}\right)\left(0.400 \text{ m}\right)}{\left(4\pi \times 10^{-7} \text{ T} \cdot \text{m} / \text{A}\right)\left(1000\right)}$$

Caution! If you use your calculator, it may not understand the keystrokes:

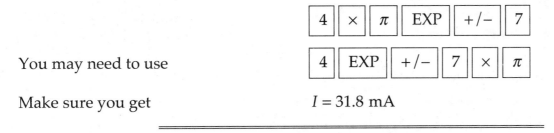

You may need to use

Make sure you get $\qquad$ $I = 31.8$ mA $\qquad \Diamond$

---

51. A positive charge $q = 3.20 \times 10^{-19}$ C moves with a velocity $\mathbf{v} = (2\mathbf{i} + 3\mathbf{j} - \mathbf{k})$ m/s through a region in which both a uniform magnetic field and a uniform electric field exist. (a) What is the total force on the moving charge (in unit-vector notation) if $\mathbf{B} = (2\mathbf{i} + 4\mathbf{j} + \mathbf{k})$ T and $\mathbf{E} = (4\mathbf{i} - \mathbf{j} - 2\mathbf{k})$ V/m? (b) What angle does the force vector make with the positive x axis?

Solution   The total force is the Lorentz force,

(a) $\mathbf{F} = q\mathbf{E} + q(\mathbf{v} \times \mathbf{B}) = q(\mathbf{E} + \mathbf{v} \times \mathbf{B})$

$\mathbf{F} = q\left[(4\mathbf{i} - \mathbf{j} - 2\mathbf{k}) \text{ V/m} + (2\mathbf{i} + 3\mathbf{j} - \mathbf{k}) \text{ m/s} \times (2\mathbf{i} + 4\mathbf{j} + \mathbf{k}) \text{ T}\right]$

$\mathbf{F} = q\left[(4\mathbf{i} - \mathbf{j} - 2\mathbf{k}) \text{ V/m} + (7\mathbf{i} - 4\mathbf{j} + 2\mathbf{k}) \text{ T} \cdot \text{m/s}\right]$

$\mathbf{F} = q\left[(11\mathbf{i} - 5\mathbf{j}) \text{ V/m}\right] = q\left[(11\mathbf{i} - 5\mathbf{j}) \text{ N/C}\right]$

$\mathbf{F} = \left(3.20 \times 10^{-19} \text{ C}\right)\left[(11\mathbf{i} - 5\mathbf{j}) \text{ N/C}\right] = (3.52\mathbf{i} - 1.60\mathbf{j}) \times 10^{-18} \text{ N} \qquad \Diamond$

(b) $\mathbf{F} \cdot \mathbf{i} = F \cos\theta = F_x \qquad\qquad \theta = \cos^{-1}\left(\frac{F_x}{F}\right) = \cos^{-1}\left(\frac{3.52}{3.87}\right) = 24.4° \qquad \Diamond$

143

65. A very long, thin strip of metal of width w carries a current I along its length, as shown in Figure P30.65. Find the magnetic field at point P in the diagram. Point P is in the plane of the strip at a distance b away from the strip.

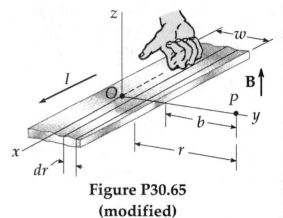

**Figure P30.65
(modified)**

Solution

Consider a long filament of the strip, which has width dr and is a distance r from point P. The magnetic field at a distance r from a long conductor is

$$B = \frac{\mu_0 I}{2\pi r}$$

Thus, the field due to the thin filament is

$$d\mathbf{B} = \frac{\mu_0 \, dI}{2\pi r}\mathbf{k} \qquad \text{where} \qquad dI = I\left(\frac{dr}{w}\right)$$

so

$$\mathbf{B} = \int_b^{b+w} \frac{\mu_0}{2\pi r}\left(I\frac{dr}{w}\right)\mathbf{k} = \frac{\mu_0 I}{2\pi w}\int_b^{b+w}\frac{dr}{r}\mathbf{k} = \frac{\mu_0 I}{2\pi w}\ln\left(1+\frac{w}{b}\right)\mathbf{k} \qquad \Diamond$$

---

63) $B = \dfrac{\mu_0 I R^2}{2(R^2+R^2)^{3/2}} = \dfrac{\mu_0 I}{2^{5/2}R}$

$I = \dfrac{2^{5/2}BR}{\mu_0} = 2^{5/2}(7\times10^{-5}\,T)(6.37\times10^6\,m)$

$I = 2.01\times10^7\,A \qquad 4\pi\times10^{-7}$

Chapter 23

Faraday's Law and Inductance

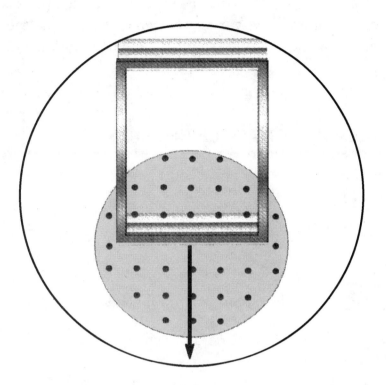

INTRODUCTION

Our studies so far have been concerned with electric fields produced by all electric charges and magnetic fields produced by moving electric charges. This chapter deals with electric fields that originate from changing magnetic fields.

Experiments conducted by Michael Faraday in England in 1831 and independently that same year by Joseph Henry in the United States showed that an electric current could be induced in a circuit by a changing magnetic field. The results of these experiments led to a basic and important law of electromagnetism known as Faraday's law of induction. This law says that the magnitude of the emf induced in a circuit equals the time rate of change of the magnetic flux through the circuit.

As we shall see, an induced emf can be produced in several ways. For instance, an induced emf and an induced current can be produced in a closed loop of wire when the wire moves into a magnetic field. We shall describe such experiments along with a number of important applications that make use of the phenomenon of electromagnetic induction.

We also describe an effect known as **self-induction,** in which a time-varying current in a conductor induces in the conductor an emf that opposes the change in the external emf that set up the current. Self-induction is the basis of the **inductor**, a circuit element that plays an important role in circuits that use time-varying currents. We discuss the energy stored in the magnetic field of an inductor and the energy density associated with a magnetic field.

NOTES FROM SELECTED CHAPTER SECTIONS

23.1 Faraday's Law of Induction

The emf induced in a circuit is equal to the time rate of change of magnetic flux through the circuit.

An emf can be induced in the circuit in several ways:

- The magnitude of the magnetic field can change as a function of time.
- The area of the circuit can change with time.
- The direction of the magnetic field relative to the circuit can change with time.
- Any combination of the above can change.

23.2 Motional emf

A potential difference, ΔV, will be maintained across a conductor moving in a magnetic field. The potential difference will be zero when the direction of motion of the conductor is parallel or antiparallel to the field direction. If the motion is reversed, the polarity of the potential difference will also be reversed.

23.3 Lenz's Law

The polarity of the induced emf is such that it tends to produce a current that will create a magnetic flux to oppose the **change in flux** through the circuit.

23.5 Self-Inductance

The self-induced emf is always proportional to the **time rate of change** of current in the circuit.

The **inductance** of a device (an inductor) depends on its geometry.

23.6 *RL Circuits*

If a resistor and an inductor are connected in series to a battery, the current in the circuit will reach an **equilibrium** value ($\mathcal{E}/R$) after a time which is long compared to the **time constant** of the circuit, $\tau = L/R$.

23.7 Energy Stored in a Magnetic Field

In an *RL* circuit, the rate at which energy is supplied by the battery equals the sum of the rate at which energy is delivered to the resistor and the rate at which energy is stored in the inductor. **The energy density is proportional to the square of the magnetic field.**

EQUATIONS AND CONCEPTS

The total magnetic flux through a plane area, A, placed in a uniform magnetic field depends on the angle between the direction of the magnetic field and the direction perpendicular to the surface area.

$$\Phi_B \equiv B_\perp A = BA\cos\theta$$

$$\Phi_{max} = BA$$

The maximum flux through the area occurs when the magnetic field is perpendicular to the plane of the surface (i.e. parallel to the surface vector, **A**). When the magnetic field is parallel to the plane of the surface (i.e. perpendicular to the area vector, **A**), the flux through the area is zero. The unit of magnetic flux is the weber, Wb.

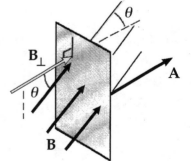

The magnetic flux through an element of surface area $d\mathbf{A}$ is given by $d\Phi_B = \mathbf{B} \cdot d\mathbf{A}$. The total magnetic flux threading a circuit is the integral of the normal component of the magnetic field over the area bounded by the circuit.

$$\Phi_B = \int \mathbf{B} \cdot d\mathbf{A} \qquad (23.1)$$

Faraday's law of induction states that the emf induced in a circuit is the rate of change of magnetic flux through the circuit. The minus sign is included to indicate the polarity of the induced emf, which can be found by use of Lenz's law.

$$\mathcal{E} = -\frac{d\Phi_B}{dt} \qquad (23.2)$$

Lenz's law states that the polarity of the induced emf in a loop (and the direction of the associated current in a closed circuit) produces a current whose magnetic field opposes the change in the flux through the loop. That is, the induced current (and the corresponding induced magnetic field) attempts to maintain the original flux through the circuit.

A "motional" emf is induced in a conductor of length ℓ, moving with speed v, perpendicular to a magnetic field.

$$\mathcal{E} = -B\ell v \qquad (23.5)$$

If the moving conductor is part of a complete circuit of resistance R, a current will be induced in the circuit.

$$I = \frac{B\ell v}{R} \qquad (23.6)$$

When a conducting coil of N turns and cross-sectional area A rotates with a constant angular velocity in a magnetic field, the emf induced in the coil will vary sinusoidally in time. For a given coil, the maximum value of the induced emf will be proportional to the angular velocity of the coil.

$$\mathcal{E} = NAB\omega\sin\omega t \qquad (23.8)$$

$$\mathcal{E}_{max} = NAB\omega$$

When the current in a coil changes in time, a self-induced emf is present in the coil. The inductance, L, is a measure of the opposition of the coil to a **change in the current**. Recall that resistance, R, is a measure of opposition to current.

$$\mathcal{E}_L = -L\frac{dI}{dt} \qquad (23.10)$$

A coil, solenoid, toroid, coaxial cable, or other conducting device is characterized by a parameter called its **inductance**, L. The inductance can be calculated, knowing the current and magnetic flux.

$$L = \frac{N\Phi_B}{I} \qquad (23.11)$$

The inductance of a given device, for example a coil, depends on its physical makeup – diameter, number of turns, type of material on which the wire is wound, and other geometric parameters. A circuit element that has a large inductance is called an inductor. The SI unit of inductance is the henry, H. A current changing at the rate of one ampere per second in a one-henry coil will generate a self-induced emf equal to one volt.

$$1\,H = 1\,\frac{V \cdot s}{A} = 1\,\Omega \cdot s$$

The **inductance** of a particular circuit element can also be expressed as the ratio of the induced emf magnitude divided by the time rate of change of current in the circuit.

$$L = -\frac{\mathcal{E}_L}{dI/dt} \qquad (23.12)$$

If the switch in the series circuit shown (which contains a battery, resistor, and inductor) is closed in position 1 at time $t = 0$, current in the circuit will increase in a characteristic fashion toward a maximum value of $(\mathcal{E}/R)$. This is shown in the graph to the right.

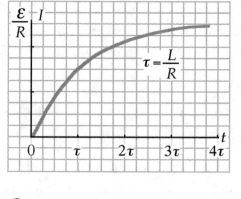

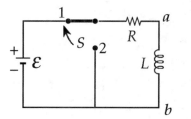

$$I(t) = \frac{\mathcal{E}}{R}(1 - e^{-t/\tau}) \qquad (23.14)$$

Let the switch in the circuit shown be at position 1 with the current at its maximum value $I_0 = \mathcal{E}/R$. If the switch is thrown to position 2 at $t = 0$, the current will decay exponentially with time. The graph at right shows the manner in which the current decays.

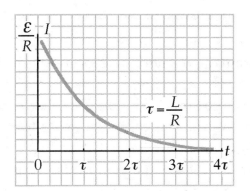

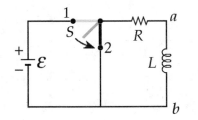

$$I(t) = \frac{\mathcal{E}}{R}e^{-t/\tau} = I_i e^{-t/\tau} \qquad (23.18)$$

$$\text{where} \quad \mathcal{E}/R = I_i$$

In Equation 23.14 and Equation 23.18, the constant in the exponent is called the **time constant** of the circuit, τ. Physically, the time constant is the time required for the current to reach 63% of its maximum value.

$$\tau = L/R \qquad (23.15)$$

The **stored energy** U_B in the magnetic field of an inductor is proportional to the square of the current in the inductor.

$$U_B = \frac{1}{2}LI^2 \qquad (23.20)$$

It is often useful to express the energy in a magnetic field as **energy density** u_B; that is, energy per unit volume.

$$u_B = \frac{B^2}{2\mu_0} \qquad (23.22)$$

SUGGESTIONS, SKILLS, AND STRATEGIES

Instantaneous and Average Induced Emf:

It is important to distinguish clearly between the **instantaneous value** of emf induced in a circuit and the **average value** of the emf induced in the circuit over a finite time interval.

To calculate the **average induced emf**, it is often useful to write Equation 23.3 as

$$\mathcal{E}_{av} = -N\left(\frac{d\Phi_B}{dt}\right)_{av} = -N\frac{\Delta\Phi_B}{\Delta t} \qquad \text{or} \qquad \mathcal{E}_{av} = -N\left(\frac{\Phi_{B,f} - \Phi_{B,i}}{\Delta t}\right)$$

where the subscripts i and f refer to the magnetic flux through the circuit at the initial and final moments of the time interval Δt. For a circuit (or coil) in a single plane, $\Phi_B = BA\cos\theta$, where θ is the angle between the direction of the normal to plane of the circuit (conducting loop) and the direction of the magnetic field.

Equation 23.4 can be used to calculate the **instantaneous value of an induced emf**. For a multiple turn coil, the induced emf is

$$\mathcal{E} = -N\frac{d}{dt}(BA\cos\theta)$$

where in a particular case B, A, θ, or any combination of those parameters can be time dependent while the others remain constant. The expression resulting from the differentiation is then evaluated using the values of B, A, and θ corresponding to the specified value.

REVIEW CHECKLIST

▷	Calculate the emf (or current) induced in a circuit when the magnetic flux through the circuit is changing in time. The variation in flux might be due to a change in (a) the area of the circuit, (b) the magnitude of the magnetic field, (c) the direction of the magnetic field, or (d) the orientation/location of the circuit in the magnetic field.

▷ Calculate the emf induced between the ends of a conducting bar as it moves through a region where there is a constant magnetic field (motional emf). Apply Lenz's law to determine the direction of an induced emf or current. You should also understand that Lenz's law is a consequence of the law of conservation of energy.

▷ Calculate the inductance of a device of suitable geometry.

▷ Calculate the magnitude and direction of the self-induced emf in a circuit containing one or more inductive elements when the current changes with time.

▷ Determine instantaneous values of the current in an RL circuit while the current is either increasing or decreasing with time.

▷ Calculate the total magnetic energy stored in a magnetic field. You should be able to perform this calculation if (1) you are given the values of the inductance of the device with which the field is associated and the current in the circuit, or (2) given the value of the magnitude of the magnetic field throughout the region of space in which the magnetic field exists. In the latter case, you must integrate the expression for the energy density u_B over an appropriate volume.

ANSWERS TO SELECTED CONCEPTUAL QUESTIONS

6. How is electric energy produced in dams? (i.e., how is the energy of motion of the water converted to ac electricity)?

Answer As the water falls, it gains kinetic energy. It is then forced to pass through a water wheel, transferring some of its energy to the rotor of a large AC electric generator.

The rotor of the generator is supplied with a small amount of DC current, which powers electromagnets in the rotor. Because the rotor is spinning, the electromagnets then create a magnetic flux that changes with time, according to the equation $\Phi_B = BA\cos\omega t$.

Coils of wire that are placed near the magnet then experience an induced emf according to the equation $\mathcal{E} = -N\,d\Phi_B/dt$.

Finally, a small amount of this electricity is used to supply the rotor with its DC current; the rest is sent out over power lines to supply customers with electricity.

In terms of energy, the hydroelectric generating station is a nonisolated system in steady state. It takes in mechanical energy and puts out a precisely equal quantity of energy, nearly all of it by electrical transmission.

8. The bar in Figure Q23.8 moves on rails to the right with a velocity **v**, and the uniform, constant magnetic field is directed out of the page. Why is the induced current clockwise? If the bar were moving to the left, what would be the direction of the induced current?

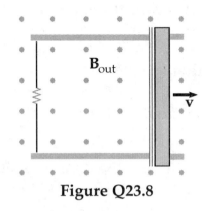

Figure Q23.8

Answer The external magnetic field is out of the paper; as the area A enclosed by the loop increases, the external flux increases according to $\Phi_B = BA\cos\theta = BA$.

As predicted by Lenz's law, due to the increase in flux through the loop, free electrons will produce a current to create a magnetic flux inside the loop to oppose the change in flux.

In this case, the magnetic field due to the current must point into the paper in order to oppose the increasing magnetic flux of the external field coming out of the paper. By the right-hand rule (with your thumb pointing in the direction of the current along each wire), the current must be clockwise.

If the bar were moving toward the left, the area would decrease, and the flux would decrease. The flux due to **B** pointing out of the paper, then, would be decreasing. In order to cancel this change, the current would have to create a magnetic field inside the loop pointing out of the paper. This time, by the right-hand rule, we see that the current must be counter-clockwise.

16. If the current in an inductor is doubled, by what factor does the stored energy change?

Answer The energy stored in an inductor carrying a current I is given by $U = \frac{1}{2}LI^2$. Therefore, doubling the current will quadruple the energy stored in the inductor.

SOLUTIONS TO SELECTED END-OF-CHAPTER PROBLEMS

3. A strong electromagnet produces a uniform magnetic field having a magnitude of 1.60 T over a cross-sectional area of 0.200 m². A coil having 200 turns and a total resistance of 20.0 Ω is placed around the electromagnet. The current is then smoothly decreased in the electromagnet until it reaches zero in 20.0 ms. What is the current induced in the coil?

Solution

The induced voltage is

$$\mathcal{E} = -N\frac{d(\mathbf{B}\cdot\mathbf{A})}{dt} = -N\left(\frac{0-B_iA\cos\theta}{\Delta t}\right)$$

Here,

$$\mathcal{E} = \frac{+200(1.60\text{ T})(0.200\text{ m}^2)(\cos 0°)}{(20.0\times 10^{-3}\text{ s})}\left(\frac{1\text{ N}\cdot\text{s}}{\text{T}\cdot\text{C}\cdot\text{m}}\right)\left(\frac{1\text{ V}\cdot\text{C}}{\text{N}\cdot\text{m}}\right) = 3200\text{ V}$$

$$I = \frac{\mathcal{E}}{R} = \frac{3200\text{ V}}{20.0\text{ }\Omega} = 160\text{ A} \qquad \Diamond$$

The positive sign means that the current in the coil flows in the same direction as the current in the electromagnet.

---

11. Figure P23.10 shows a top view of a bar that can slide without friction. The resistor is 6.00 Ω and a 2.50-T magnetic field is directed perpendicularly downward, into the paper. Let ℓ = 1.20 m. (a) Calculate the applied force required to move the bar to the right at a constant speed of 2.00 m/s. (b) At what rate is energy delivered to the resistor?

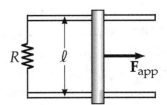

Figure P23.10

Solution

(a) We use the rigid body in equilibrium model. At constant speed, the net force on the moving bar equals zero, or $\left|\mathbf{F}_{app}\right| = I|\boldsymbol{\ell}\times\mathbf{B}|$ where the current in the bar is $I = \mathcal{E}/R$ and $\mathcal{E} = B\ell v$. Therefore,

$$F_{app} = \left(\frac{B\ell v}{R}\right)\ell B = \frac{B^2\ell^2 v}{R} = \frac{(2.50\text{ T})^2(1.20\text{ m})^2(2.00\text{ m/s})}{6.00\text{ }\Omega} = 3.00\text{ N} \qquad \Diamond$$

6) b) $\mathcal{E} = -N\int\frac{d(B\cdot A)}{dt}$

(b) $$\mathcal{P} = F_{\text{app}}v = (3.00 \text{ N})(2.00 \text{ m/s}) = 6.00 \text{ W}$$ ◊

In terms of energy, the circuit is a nonisolated system in steady state. The energy, six joules every second, delivered to the circuit by work done by the applied force is delivered to the resistor by electrical transmission. The energy can leave the resistor as energy transferred by heat into the surrounding air.

---

17. A conducting rectangular loop of mass M, resistance R, and dimensions w by ℓ falls from rest into a magnetic field **B** as in Figure P23.17. While only the bottom edge of the loop is immersed in the field, the loop approaches terminal speed v_T. (a) Show that

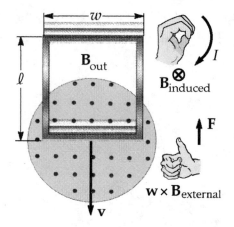

$$v_T = \frac{MgR}{B^2w^2}$$

(b) Why is v_T proportional to R? (c) Why is it inversely proportional to B^2?

Figure P23.17

Solution

Let y represent the average vertical height of the strong-field region above the bottom edge of the loop. As the loop falls, y increases; the loop encloses increasing flux toward you and has induced in it an emf to produce a current to make its own magnetic field away from you. This current is to the left in the bottom side of the loop, and feels an upward force in the external field.

(a) Symbolically, the flux is $$\Phi = BA\cos\theta = Bwy\cos 0°$$

The emf is $$\mathcal{E} = -N\frac{d}{dt}(Bwy) = -Bw\frac{dy}{dt} = -Bwv$$

The magnitude of the current is $$I = \frac{|\mathcal{E}|}{R} = \frac{Bwv}{R}$$

and the force is $$\mathbf{F}_B = I\,\mathbf{w}\times\mathbf{B} = \left(\frac{Bwv}{R}\right)wB\sin 90° = \frac{B^2w^2v}{R} \text{ up}$$

When the loop is moving at terminal speed, we may model it as a rigid body in equilibrium:

$\Sigma F_y = 0$ becomes

$$\frac{+B^2 w^2 v_T}{R} - Mg = 0$$

Thus,

$$v_T = \frac{MgR}{B^2 w^2} \qquad \Diamond$$

(b) The emf is directly proportional to v, but the current is inversely proportional to R. A large R means a small current at a given speed, so the loop must travel faster to get F_B = weight. $\Diamond$

(c) At a given speed, the current is directly proportional to the magnetic field. But the force is proportional to the product of the current and the field. For a small B, the speed must increase to compensate for both the small B and also the current, so $v_T \propto B^{-2}$. $\Diamond$

---

19. A coil of area 0.100 m^2 is rotating at 60.0 rev/s with the axis of rotation perpendicular to a 0.200-T magnetic field. (a) If the coil has 1000 turns, what is the maximum voltage generated in it? (b) What is the orientation of the coil with respect to the magnetic field when the maximum induced voltage occurs?

Solution For a coil rotating in a magnetic field,

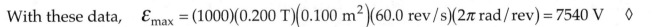

(a) By Equation 23.8, $\mathcal{E} = NBA\omega \sin\theta$

so $\mathcal{E}_{max} = NBA\omega$

With these data, $\mathcal{E}_{max} = (1000)(0.200 \text{ T})(0.100 \text{ m}^2)(60.0 \text{ rev/s})(2\pi \text{ rad/rev}) = 7540 \text{ V} \quad \Diamond$

(b) $|\mathcal{E}|$ approaches $\mathcal{E}_{max}$ when $|\sin\theta|$ approaches 1 or $\theta = \pm\dfrac{\pi}{2}$

Therefore, at maximum emf, the plane of the coil is parallel to the field. $\Diamond$

---

21. A magnetic field directed into the page changes with time according to the expression $B = (0.0300t^2 + 1.40)$ T, where t is in seconds. The field has a circular cross-section of radius $R = 2.50$ cm (Fig. P23.21). What are the magnitude and direction of the electric field at point P_1 when $t = 3.00$ s and $r_1 = 0.0200$ m?

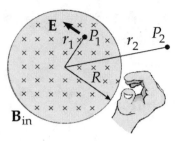

Solution

$$\oint \mathbf{E} \cdot ds = -\frac{d\Phi_B}{dt}$$

Figure P23.21

Consider a circular integration path of radius r_1

$$E(2\pi r_1) = -\frac{d}{dt}(BA) = -A\left(\frac{dB}{dt}\right)$$

$$|E| = \frac{A}{2\pi r_1}\frac{d}{dt}(0.0300t^2 + 1.40)\text{ T} = \frac{\pi r_1^2}{2\pi r_1}(0.0600t) = \frac{r_1}{2}(0.0600t)$$

At $t = 3.00$ s, $\qquad E = \left(\frac{0.0200\text{ m}}{2}\right)(0.0600\text{ T/sec})(3.00\text{ sec}) = 1.80 \times 10^{-3}\text{ N/C}$ ◊

If there were a circle of wire of radius r_1, it would enclose increasing magnetic flux due to a magnetic field away from you. It would carry counterclockwise current to make its own magnetic field toward you, to oppose the change. Even without the wire and current, the counterclockwise electric field that would cause the current is lurking. At point P_1, it is upward and to the left, perpendicular to r_1. ◊

---

25. A 10.0-mH inductor carries a current $I = I_{max}\sin\omega t$, with $I_{max} = 5.00$ A and $\omega/2\pi = 60.0$ Hz. What is the back emf as a function of time?

Solution $\qquad \mathcal{E}_{back} = -\mathcal{E}_L = L\frac{dI}{dt} = L\frac{d}{dt}(I_{max}\sin\omega t)$

$$\mathcal{E}_{back} = L\omega I_{max}\cos\omega t = (0.0100\text{ H})(120\ \text{s}^{-1})(5.00\text{ A})\cos(120\ t)$$

$$\mathcal{E}_{back} = (18.8\text{ V})\cos(377t)$$ ◊

---

31. A 12.0-V battery is connected into a series circuit containing a 10.0-Ω resistor and a 2.00-H inductor. How long will it take the current to reach (a) 50.0% and (b) 90.0% of its final value?

Solution

The time constant is $\tau = \dfrac{L}{R} = 0.200$ s:

(a) In $I = \dfrac{\mathcal{E}(1-e^{-t/\tau})}{R}$, the final value, which the current approaches, is

$$I = \frac{\mathcal{E}(1-e^{-\infty})}{R} = \frac{\mathcal{E}}{R}$$

We have at 50%,

$$(0.500)\left(\frac{\mathcal{E}}{R}\right) = \frac{\mathcal{E}(1-e^{-t/0.200\ \text{s}})}{R}$$

Solving for t,

$$0.500 = 1 - e^{-t/0.200\ \text{s}}$$

$$e^{-t/0.200\ \text{s}} = 0.500$$

$$e^{t/0.200\ \text{s}} = 2.00$$

$$\frac{t}{0.200\ \text{s}} = \ln 2.00 = 0.693$$

Thus,

$$t = 0.139\ \text{s} \qquad \Diamond$$

(b) At 90%,

$$0.900 = 1 - e^{-t/\tau}$$

and

$$t = \tau \ln\left(\frac{1}{1-0.900}\right)$$

$$t = (0.200\ \text{s})\ln 10.0 = 0.461\ \text{sec} \qquad \Diamond$$

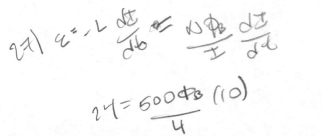

$36) \quad 1.5 - (1.25)e^{-10t/5}$

37. A 140-mH inductor and a 4.90-Ω resistor are connected with a switch to a 6.00-V battery as shown in Figure P23.37. (a) If the switch is thrown to the left (connecting the battery), how much time elapses before the current reaches 220 mA? (b) What is the current in the inductor 10.0 s after the switch is closed? (c) Now the switch is quickly thrown from A to B. How much time elapses before the current falls to 160 mA?

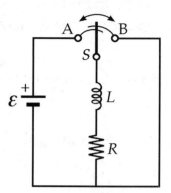

Figure P23.37

Solution   The general LR equation is obtained by combining equations 23.14 and 23.15:

(a)
$$I = \frac{\mathcal{E}\left(1 - e^{-Rt/L}\right)}{R}$$

or
$$0.220 \text{ A} = \frac{6.00 \text{ V}}{4.90 \text{ }\Omega}\left(1 - e^{-(4.90 \text{ }\Omega)t/0.140 \text{ H}}\right)$$

$$0.180 = 1 - e^{-\left(35.0 \text{ s}^{-1}\right)t}$$

or
$$e^{\left(35.0 \text{ s}^{-1}\right)t} = 1.22$$

Thus, $t = \dfrac{\ln 1.22}{35.0 \text{ s}^{-1}} = 5.66 \text{ ms}$ ◊

(b)   Again referring to the general equation,

$$I = \frac{6.00 \text{ V}}{4.90 \text{ }\Omega}\left(1 - e^{\left(-35.0 \text{ s}^{-1}\right)(10.0 \text{ s})}\right) = 1.22 \text{ A}$$ ◊

(c)
$$0.160 \text{ A} = (1.22 \text{ A})e^{(-4.90 \text{ }\Omega)t/0.140 \text{ H}}$$

$$7.65 = e^{\left(35.0 \text{ s}^{-1}\right)t} \qquad \text{and} \qquad t = \frac{\ln 7.65}{35.0 \text{ s}^{-1}} = 58.1 \text{ ms}$$ ◊

39. An air-core solenoid with 68 turns is 8.00 cm long and has a diameter of 1.20 cm. How much energy is stored in its magnetic field when it carries a current of 0.770 A?

Solution   For a solenoid of length ℓ,

$$L = \frac{\mu_0 N^2 A}{\ell}$$

Thus, since

$$U_B = \frac{1}{2}LI^2 = \frac{\mu_0 N^2 A I^2}{2\ell}$$

$$U_B = \frac{\left(4\pi \times 10^{-7}\ \text{N}/\text{A}^2\right)(68)^2 \pi \left(6.00 \times 10^{-3}\ \text{m}\right)^2 (0.770\ \text{A})^2}{2(0.0800\ \text{m})} = 2.44 \times 10^{-6}\ \text{J} \qquad \Diamond$$

41. On a clear day at a certain location, a 100-V/m vertical electric field exists near the Earth's surface. At the same place, the Earth's magnetic field has a magnitude of 0.500×10^{-4} T. Compute the energy density of the two fields.

Solution

$$u_E = \frac{\epsilon_0 E^2}{2} = \frac{\left(8.85 \times 10^{-12}\ \text{C}^2/\text{N} \cdot \text{m}^2\right)(100\ \text{N}/\text{C})^2 (1\ \text{J}/\text{N} \cdot \text{m})}{2} = 44.2\ \text{nJ}/\text{m}^3 \qquad \Diamond$$

$$u_B = \frac{B^2}{2\mu_0} = \frac{\left(5.00 \times 10^{-5}\ \text{T}\right)^2}{2\left(4\pi \times 10^{-7}\ \text{T} \cdot \text{m}/\text{A}\right)} = 995 \times 10^{-6}\ \text{T} \cdot \text{A}/\text{m}$$

$$u_B = \left(995 \times 10^{-6}\ \text{T} \cdot \text{A}/\text{m}\right)(1\ \text{N} \cdot \text{s}/\text{T} \cdot \text{C} \cdot \text{m})(1\ \text{J}/\text{N} \cdot \text{m}) = 995\ \mu\text{J}/\text{m}^3 \qquad \Diamond$$

Magnetic energy density is 22500 times greater than that in the electric field.

47. The magnetic flux through a metal ring varies with time t according to $\Phi_B = 3(at^3 - bt^2)$ T·m², with $a = 2.00$ s⁻³ and $b = 6.00$ s⁻². The resistance of the ring is 3.00 Ω. Determine the maximum current induced in the ring during the interval from $t = 0$ to $t = 2.00$ s.

Solution

Substituting the given values,
$$\Phi_B = \left(6.00t^3 - 18.0t^2\right) \text{T·m}^2$$

Therefore, the emf induced is
$$\mathcal{E} = -\frac{d\Phi_B}{dt} = -18.0t^2 + 36.0t$$

The maximum $\mathcal{E}$ occurs when
$$\frac{d\mathcal{E}}{dt} = -36.0t + 36.0 = 0, \text{ which gives } t = 1.00 \text{ s}.$$

Thus, maximum current (at $t = 1.00$ s) is $I_{max} = \dfrac{\mathcal{E}}{R} = \dfrac{(-18.0 + 36.0) \text{ V}}{3.00 \ \Omega} = 6.00$ A   ◊

---

55. A long, straight wire carries a current $I = I_{max} \sin(\omega t + \phi)$ and lies in the plane of a rectangular coil of N turns of wire, as shown in Figure P23.6. The quantities I_{max}, ω, and ϕ are all constants. Determine the emf induced in the coil by the magnetic field created by the current in the straight wire. Assume $I_{max} = 50.0$ A, $\omega = 200\pi$ s⁻¹, $N = 100$, $h = \omega = 5.00$ cm, and $L = 20.0$ cm.

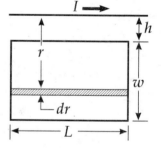

Figure P23.6 (modified)

Solution   The coil is the boundary of a rectangular area. The magnetic field produced by the current in the straight wire is perpendicular to the plane of the area at all points. The magnitude of the field is

$$B = \frac{\mu_0 I}{2\pi r}$$

Thus the flux is
$$N\Phi_B = \frac{\mu_0 NL}{2\pi} I \int_h^{h+w} \frac{dr}{r} = \frac{\mu_0 NL}{2\pi} I_{max} \ln\left(\frac{h+w}{h}\right) \sin(\omega t + \phi)$$

Finally, the induced emf is in absolute value

$$|\mathcal{E}| = N\frac{d\Phi_B}{dt} = \frac{\mu_0 NL}{2\pi}I_{max}\omega\ln\left(\frac{h+w}{h}\right)\cos(\omega\phi+\ )$$

$$|\mathcal{E}| = \left(4\pi\times10^{-7}\ \text{T}\cdot\text{m}/\text{A}\right)\frac{(100)(0.200\ \text{m})}{2\pi}(50.0\ \text{A})(200\pi\ \text{s}^{-1})\ln\left(\frac{10.0\ \text{cm}}{5.00\ \text{cm}}\right)\cos(\omega t\ \phi\ )$$

$$|\mathcal{E}| = (87.1\ \text{mV})\cos(200\pi t + \phi) \qquad \lozenge$$

Related Comments   The term $\sin(\omega t + \phi)$ in the expression for the current in the straight wire does not change appreciably when ωt changes by 0.1 rad or less. Thus, the current does not change appreciably during a time interval

$$\Delta t < \frac{0.100}{200\pi\ \text{s}^{-1}} = 1.59\times10^{-4}\ \text{s}$$

We define a critical length,

$$c\Delta t = \left(3.00\times10^8\ \text{m}/\text{s}\right)\left(1.59\times10^{-4}\ \text{s}\right) = 4.77\times10^4\ \text{m}$$

equal to the distance to which field changes could be propagated during an interval of 1.59×10^{-4} s. This length is so much larger than any dimension of the loop or its distance from the wire that, although we consider the straight wire to be infinitely long, we can also safely ignore the field propagation effects in the vicinity of the loop. Moreover, the phase angle can be considered to be constant along the wire in the vicinity of the loop. If the frequency ω were much larger, say $200\pi\times10^5$ s^{-1}, the corresponding critical length would be only 48 cm. In this situation, propagation effects would be important and the above expression of $\mathcal{E}$ would require modification. As a "rule of thumb," we can consider field propagation effects for circuits of laboratory size to be negligible for frequencies, $f = \omega/2\pi$, that are less than about 10^6 Hz.

61) a) $B = \dfrac{\left(4\pi\times10^{7}\right)\left(5\times10^3 A\right)}{2\pi\ (.05\,m)} = .057$

b) $B = \dfrac{4\pi\times10^{7}\ (5\times10^3)}{2\pi\ (.07)} = .02T$

c) $U = \dfrac{4\pi\times10^{7}\ (5\times10^3)^2\ (100)}{4\pi}$

$\cdot\ln\left(\frac{5}{2}\right) = 2.29\times10$

Chapter 24

Electromagnetic Waves

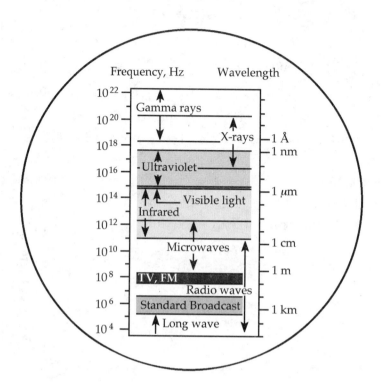

INTRODUCTION

By definition, mechanical disturbances, such as sound waves, water waves, and waves on a string, require the presence of a medium. This chapter is concerned with the properties of electromagnetic waves that (unlike mechanical waves) can propagate through empty space.

The consequences of Maxwell's equations are far-reaching and very dramatic for the history of physics. One of them, the Ampere-Maxwell law, predicts that a time-varying electric field produces a magnetic field just as a time-varying magnetic field produces an electric field (Faraday's law). From this generalization, Maxwell introduced the concept of displacement current, a new source of a magnetic field. Thus, Maxwell's theory provided the final important link between electric and magnetic fields.

NOTES FROM SELECTED CHAPTER SECTIONS

24.2   Maxwell's Equations

Electromagnetic waves are generated by accelerating electric charges. The radiated waves consist of oscillating electric and magnetic fields, which are **at right angles to each other** and also **at right angles to the direction of wave propagation.**

The fundamental laws describing the behavior of electric and magnetic fields are Maxwell's equations. In this unified theory of electromagnetism, Maxwell showed that electromagnetic waves are a natural consequence of these fundamental laws.

The theory he developed is based upon the following pieces of information:

- A charge creates an electric field. Electric field lines originate on positive charges and terminate on negative charges. The relationship between charges and the fields they produce is described by Gauss's law.

- Magnetic field lines always form closed loops; that is, they do not begin or end anywhere.

- A varying magnetic field induces an emf and hence an electric field. This is a statement of Faraday's law.

- A moving charge (constituting a current) creates a magnetic field, as summarized in Ampère's law.

- A varying electric field creates a magnetic field. This is Maxwell's addition to Ampère's law.

24.3   Electromagnetic Waves

Following is a summary of the properties of electromagnetic waves:

- The solutions of Maxwell's third and fourth equations are wavelike, where both **E** and **B** satisfy the same wave equation.

- Electromagnetic waves travel through empty space with the speed of light,

$$c = \frac{1}{\sqrt{\epsilon_0 \mu_0}}$$

- The electric and magnetic field components of plane electromagnetic waves are perpendicular to each other and also perpendicular to the direction of wave propagation. The latter property can be summarized by saying that electromagnetic waves are transverse waves.

- The magnitudes of **E** and **B** in empty space are related by $E/B = c$.

- Electromagnetic waves obey the principle of superposition.

24.5   Energy Carried by Electromagnetic Waves

The magnitude of the Poynting vector represents the rate at which energy flows through a unit surface area perpendicular to the flow.

For an electromagnetic wave, the instantaneous energy density associated with the magnetic field equals the instantaneous energy density associated with the electric field. Hence, in a given volume, the energy is equally shared by the two fields.

24.6   Momentum and Radiation Pressure

Electromagnetic waves have momentum and exert pressure on surfaces on which they are incident. The pressure exerted by a normally incident wave on a **totally reflecting** surface is **double** that exerted on a surface which **completely absorbs** the incident wave.

EQUATIONS AND CONCEPTS

Maxwell's equations are the fundamental laws governing the behavior of electric and magnetic fields. Electromagnetic waves are a natural consequence of these laws.

Gauss's law: The total electric flux through any closed surface equals the net charge enclosed by the surface divided by ϵ_0.

$$\oint \mathbf{E} \cdot d\mathbf{A} = \frac{Q}{\epsilon_0} \tag{24.4}$$

Gauss's law for magnetism: The net magnet flux through a closed surface is zero.

$$\oint \mathbf{B} \cdot d\mathbf{A} = 0 \tag{24.5}$$

Faraday's law of induction: The line integral of the electric field around any closed path equals the rate of change of magnetic flux through any surface area bounded by the path.

$$\oint \mathbf{E} \cdot d\mathbf{s} = -\frac{d\Phi_B}{dt} \tag{24.6}$$

$$\text{where} \quad \Phi_B = \mathbf{B} \cdot \mathbf{A}$$

Generalized form of Ampere's law: The line integral of the magnetic field around any closed path is determined by the net current and the rate of change of electric flux through any surface bounded by the path.

$$\oint \mathbf{B} \cdot d\mathbf{s} = \mu_0 I + \epsilon_0 \mu_0 \frac{d\Phi_E}{dt} \tag{24.7}$$

$$\text{where} \quad \Phi_E = \mathbf{E} \cdot \mathbf{A}$$

You should notice that the integrals in Equations 24.4 and 24.5 are **surface integrals** in which the normal components of electric and magnetic fields are integrated over a **closed surface** while Equations 24.6 and 24.7 involve line integrals in which the tangential components of electric and magnetic fields are integrated around a **closed path**.

Both **E** and **B** satisfy a differential equation, which has the form of the general wave equation. These are the wave equations for electromagnetic waves in free space (where $Q = 0$ and $I = 0$). As stated here, they represent linearly polarized waves traveling with a speed c.

$$\frac{\partial^2 E}{\partial x^2} = \epsilon_0 \mu_0 \frac{\partial^2 E}{\partial t^2} \tag{24.15}$$

$$\frac{\partial^2 B}{\partial x^2} = \epsilon_0 \mu_0 \frac{\partial^2 B}{\partial t^2} \tag{24.16}$$

$$c = \frac{1}{\sqrt{\epsilon_0 \mu_0}} \tag{24.17}$$

The electric and magnetic fields vary in position and time as **sinusoidal transverse waves**. Their planes of vibration are perpendicular to each other and perpendicular to the direction of propagation.

$$E = E_{max} \cos(kx - \omega t) \tag{24.18}$$

$$B = B_{max} \cos(kx - \omega t) \tag{24.19}$$

The ratio of the magnitude of the electric field to the magnitude of the magnetic field is constant and equal to the speed of light c.

$$\frac{E}{B} = c \tag{24.20}$$

The **Poynting vector S** describes the energy flow associated with an electromagnetic wave. The direction of **S** is along the direction of propagation and the magnitude of **S** is the rate at which electromagnetic energy crosses a unit surface area perpendicular to the direction of **S**.

$$\mathbf{S} \equiv \frac{1}{\mu_0} \mathbf{E} \times \mathbf{B} \tag{24.22}$$

The **wave intensity** is the time average of the magnitude of the Poynting vector. E_{max} and B_{max} are the **maximum values** of the field magnitudes.

$$I = S_{av} = \frac{E_{max}^{2}}{2\mu_0 c} = \frac{c B_{max}^{2}}{2\mu_0} \tag{24.24}$$

The electric and magnetic fields have **equal instantaneous energy densities.**

$$u_B = u_E$$

$$u_E = \frac{1}{2}\epsilon_0 E^2 \qquad (24.25)$$

$$u_B = \frac{B^2}{2\mu_0} \qquad (24.26)$$

The total instantaneous energy density u is proportional to E^2 and to B^2 while the **total average energy density** is proportional to E_{max}^2 and to B_{max}^2. The average energy density is also proportional to the wave intensity.

$$u = \epsilon_0 E^2 = \frac{B^2}{\mu_0}$$

$$u_{av} = \frac{1}{2}\epsilon_0 E_{max}^2 = \frac{B_{max}^2}{2\mu_0} \qquad (24.27)$$

$$I = S_{av} = c\,u_{av} \qquad (24.28)$$

The **linear momentum p delivered to an absorbing surface** by an electromagnetic wave at normal incidence depends on the fraction of the total energy absorbed.

$$p = \frac{U}{c} \text{ (complete absorption)} \qquad (24.29)$$

$$p = \frac{2U}{c} \text{ (complete reflection)}$$

An absorbing surface (at normal incidence) will experience a **radiation pressure P**, which depends on the intensity of the wave (power per unit area) and the degree of absorption.

$$P = \frac{I}{c} \qquad \text{(complete absorption)} \qquad (24.30)$$

$$P = \frac{2I}{c} \qquad \text{(complete reflection)}$$

REVIEW CHECKLIST

▷ Describe the essential features of the apparatus and procedure used by Hertz in his experiments leading to the discovery and understanding of the source and nature of electromagnetic waves.

▷ For a properly described plane electromagnetic wave, calculate the values for the Poynting vector (magnitude), wave intensity, and instantaneous and average energy densities.

▷ Calculate the radiation pressure on a surface and the linear momentum delivered to a surface by an electromagnetic wave.

▷ Understand the production of electromagnetic waves and radiation of energy by an oscillating dipole. Use a diagram to show the relative directions for **E**, **B**, and **S**.

ANSWERS TO SELECTED CONCEPTUAL QUESTIONS

6. If you charge a comb by running it through your hair and then hold the comb next to a bar magnet, do the electric and magnetic fields that are produced constitute an electromagnetic wave?

Answer

No. Charge on the comb creates an electric field and the bar magnet sets up a magnetic field. However, at any point these fields are constant in magnitude and direction. In an electromagnetic wave, the electric and magnetic fields must be changing in order to create each other.

□   □   □   □

10. Suppose an alien from another planet had eyes that were sensitive to infrared radiation. Describe what the alien would see if he looked around the room you are now in. That is, what would be bright, and what would be dim?

Answer

Light bulbs and the toaster glow brightly in the infrared. Somewhat fainter are the back of the refrigerator and the back of the television set, while the TV screen is dark. The pipes under the sink show the same weak glow as the walls until you turn on the faucets. Then the pipe on the right gets darker while that on the left develops a rich gleam that quickly runs up along its length. The food on your plate shines; so does human skin, the same color for all races. Clothing is dark as a rule, but your seat glows like a monkey's rump when you get up from a chair and you leave a patch of the same glow on your chair. Your face appears lit from within, like a jack-o-lantern; your nostrils and openings of your ear canals are bright; brighter still are the pupils of your eyes.

13. Radio stations often advertise "instant news." If what they mean is that you hear the news at the instant they speak it, is their claim true? About how long would it take for a message to travel across this country by radio waves, assuming that these waves could travel this great distance and still be detected?

Answer

Radio waves move at the speed of light. They can travel around the curved surface of the Earth, bouncing between the ground and the ionosphere, which has an altitude that is small when compared to the radius of the earth. The distance across the lower forty-eight states is approximately 5000 km, requiring a time of $(5 \times 10^6 \text{ m})/(3 \times 10^8 \text{ m/s}) \sim 10^{-2}$ s. To go halfway around the Earth takes only 0.07 s. In other words, a speech can be heard on the other side of the world before it is heard at the back of the room.

SOLUTIONS TO SELECTED END-OF-CHAPTER PROBLEMS

5.   Figure 24.6 shows a plane electromagnetic sinusoidal wave propagating in the x direction. Suppose that the wavelength is 50.0 m and the electric field vibrates in the xy plane with an amplitude of 22.0 V/m. Calculate (a) the frequency of the wave and (b) the magnitude and direction of **B** when the electric field has its maximum value in the negative y direction. (c) Write an expression for B in the form

$$B = B_{max} \cos(kx - \omega t)$$

with numerical values for B_{max}, k, and ω.

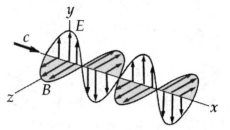

Figure 24.6

Solution

(a) $c = f\lambda$ $\qquad\qquad f = \dfrac{c}{\lambda} = \dfrac{3.00 \times 10^8 \text{ m/s}}{50.0 \text{ m}} = 6.00 \times 10^6 \text{ Hz}$ ◊

(b) $c = \dfrac{E}{B}$ $\qquad\qquad B = \dfrac{E}{c} = \dfrac{22.0 \text{ V/m}}{3.00 \times 10^8 \text{ m/s}} = 7.33 \times 10^{-8} \text{ T} = 73.3 \text{ nT}$ ◊

   B is directed along **negative z direction** when **E** is in the negative y direction; therefore, $\mathbf{S} = \mathbf{E} \times \mathbf{B}/\mu_0$ will propagate in the direction of $(-\mathbf{j}) \times (-\mathbf{k}) = +\mathbf{i}$.

(c) $B = B_{max} \cos(kx - \omega t)$

   $k = \dfrac{2\pi}{\lambda} = \dfrac{2\pi}{50.0 \text{ m}} = 0.126 \text{ m}^{-1}$

   $\omega = 2\pi f = (2\pi \text{ rad})(6.00 \times 10^6 \text{ Hz}) = 3.77 \times 10^7 \text{ rad/s}$

   Thus, $B = (73.3 \text{ nT}) \cos\left[(0.126 \text{ rad/m})x - (3.77 \times 10^7 \text{ rad/s})t\right]$ ◊

13. A fixed inductance $L = 1.05~\mu H$ is used in series with a variable capacitor in the tuning section of a radio. What capacitance tunes the circuit to the signal from a transmitter broadcasting at 6.30 MHz?

Solution It is difficult to predict a value for the capacitance without doing the calculations, but we might expect a typical value to be in the μF or pF range.

We want the resonance frequency of the circuit to match the broadcasting frequency, and for a simple *RLC* circuit, the resonance frequency only depends on the magnitudes of the inductance and capacitance.

The resonance frequency is
$$f_0 = \frac{1}{2\pi\sqrt{LC}}$$

Thus,
$$C = \frac{1}{(2\pi f_0)^2 L} = \frac{1}{\left[2\pi\left(6.30\times10^6~\text{Hz}\right)\right]^2\left(1.05\times10^{-6}~\text{H}\right)} = 608~\text{pF} \qquad \Diamond$$

This is indeed a typical capacitance, so our calculation appears reasonable. However, you probably would not hear any familiar music on this broadcasting frequency. The frequency range for FM radio broadcasting is 88.0 – 108.0 MHz, and AM radio is 535 – 1605 kHz. The 6.30 MHz frequency falls in the Maritime Mobile SSB Radiotelephone range, so you might hear a ship captain instead of Top 40 tunes!

This and other information about the radio frequency spectrum can be found on the National Telecommunications and Information Administration (NTIA) website, which at the time of this printing was at http://www.ntia.doc.gov/osmhome/allochrt.html .

19. A community plans to build a facility to convert solar radiation to electric power. They require 1.00 MW of power, and the system to be installed has an efficiency of 30.0% (i.e., 30.0% of the solar energy incident on the surface is converted to electric energy). What must be the effective area of a perfectly absorbing surface used in such an installation, assuming a constant intensity of 1000 W/m^2?

Solution At 30.0% efficiency, the power $\mathcal{P} = 0.300\,SA$:

$$A = \frac{\mathcal{P}}{0.300\,S} = \frac{1.00\times10^6~\text{W}}{0.300\left(1000~\text{W}/\text{m}^2\right)} = 3330~\text{m}^2 \cong 0.75~\text{acres} \qquad \Diamond$$

21. The filament of an incandescent lamp has a 150-Ω resistance and carries a direct current of 1.00 A. The filament is 8.00 cm long and 0.900 mm in radius. (a) Calculate the magnitude of the Poynting vector at the surface of the filament. (b) Find the magnitude of the electric and magnetic fields at the surface of the filament.

Solution

In this problem, the Poynting vector does not represent the light radiated or the energy convected away from the filament. Rather, it represents the energy flow in the static electric and magnetic fields created in the surrounding empty space, by current in the filament. The Poynting vector describes not only energy transfer by electromagnetic radiation but also energy transfer by electrical transmission.

The rate at which energy is delivered to the resistor is

$$\mathcal{P} = I^2 R = (1.00 \text{ A})^2 (150 \ \Omega) = 150 \text{ W}$$

and the surface area is $\quad A = 2\pi r L = 2\pi (0.900 \times 10^{-3} \text{ m})(0.0800 \text{ m}) = 4.52 \times 10^{-4} \text{ m}^2$.

(a) The Poynting vector is directed radially inward:

$$S = \frac{\mathcal{P}}{A} = \frac{150 \text{ W}}{4.52 \times 10^{-4} \text{ m}^2} = 3.32 \times 10^5 \text{ W/m}^2 \qquad \lozenge$$

(b) $\quad B = \mu_0 \dfrac{I}{2\pi r} = \left(4\pi \times 10^{-7} \text{ T} \cdot \text{m/A}\right) \dfrac{1.00 \text{ A}}{2\pi (0.900 \times 10^{-3} \text{ m})} = 2.22 \times 10^{-4} \text{ T} \qquad \lozenge$

$$E = \frac{\Delta V}{\Delta x} = \frac{IR}{L} = \frac{150 \text{ V}}{0.0800 \text{ m}} = 1880 \text{ V/m} \qquad \lozenge$$

Note: We could also calculate the Poynting vector from $S = \dfrac{EB}{\mu_0} = 3.32 \times 10^5 \text{ W/m}^2 \qquad \lozenge$

25. A radio wave transmits 25.0 W/m² of power per unit area. A flat surface of area *A* is perpendicular to the direction of propagation of the wave. Calculate the radiation pressure on it if the surface is a perfect absorber.

Solution For complete absorption,

$$P = \frac{S}{c} = \frac{25.0 \text{ W/m}^2}{3.00 \times 10^8 \text{ m/s}} = 8.33 \times 10^{-8} \text{ N/m}^2 \qquad \Diamond$$

27. A 15.0-mW helium-neon laser ($\lambda = 632.8$ nm) emits a beam of circular cross-section with a diameter of 2.00 mm. (a) Find the maximum electric field in the beam. (b) What total energy is contained in a 1.00-m length of the beam? (c) Find the momentum carried by a 1.00-m length of the beam.

Solution The intensity of the light is the average magnitude of the Poynting vector:

$$I = \frac{\mathcal{P}}{\pi r^2} = \frac{E_{max}^2}{2\mu_0 c}$$

(a) Therefore, the maximum electric field is

$$E_{max} = \sqrt{\frac{\mathcal{P}(2\mu_0 c)}{\pi r^2}} = 1.90 \times 10^3 \text{ N/C} \qquad \Diamond$$

(b) The power being 15.0 mW means that 15.0 mJ passes through a cross section of the beam in one second. This energy is uniformly spread through a beam length of 3.00×10^8 m, since that is how far the front end of the energy travels in one second. Thus, the energy in just a one-meter length is

$$\left(\frac{15.0 \times 10^{-3} \text{ J/s}}{3.00 \times 10^8 \text{ m/s}}\right)(1.00 \text{ m}) = 5.00 \times 10^{-11} \text{ J} \qquad \Diamond$$

(c) The linear momentum carried by a 1.00-m length of the beam is the momentum that would be received by an absorbing surface, under complete absorption:

$$p = \frac{U}{c} = \frac{5.00 \times 10^{-11} \text{ J}}{3.00 \times 10^8 \text{ m/s}} = 1.67 \times 10^{-19} \text{ kg} \cdot \text{m/s} \qquad \Diamond$$

33. What wavelengths of electromagnetic waves in free space have frequencies of (a) 5.00×10^{19} Hz and (b) 4.00×10^{9} Hz?

Solution

(a) $\lambda = \dfrac{c}{f} = \dfrac{3.00 \times 10^{8} \text{ m/s}}{5.00 \times 10^{19} \text{ s}^{-1}} = 6.00 \text{ pm}$ ◊

(b) $\lambda = \dfrac{c}{f} = \dfrac{3.00 \times 10^{8} \text{ m/s}}{4.00 \times 10^{9} \text{ s}^{-1}} = 7.50 \text{ cm}$ ◊

The electromagnetic wave we found in part (a) would be called an x-ray if it were emitted when an inner electron in an atom loses energy, or an electron in a vacuum tube. It would be called a gamma ray if it were radiated by an atomic nucleus.

The finger-length electromagnetic wave we found in part (b) is called a radio wave or a microwave.

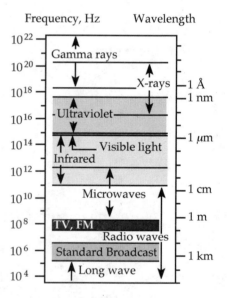

Figure 24.13

37. Plane-polarized light is incident on a single polarizing disk with the direction of E_0 parallel to the direction of the transmission axis. Through what angle should the disk be rotated so that the intensity in the transmitted beam is reduced by a factor of (a) 3.00, (b) 5.00, (c) 10.0?

Solution   We define the initial angle, at which all the light is transmitted to be $\theta = 0$. Turning the disk to another angle will then reduce the transmitted light by an intensity factor of $I = I_0 \cos^2 \theta$.

(a)   For $I = \dfrac{I_0}{3.00}$, $\cos\theta = \dfrac{1}{\sqrt{3.00}}$ and $\theta = 54.7°$ ◊

(b)   For $I = \dfrac{I_0}{5.00}$, $\cos\theta = \dfrac{1}{\sqrt{5.00}}$ and $\theta = 63.4°$ ◊

(c)   For $I = \dfrac{I_0}{10.0}$, $\cos\theta = \dfrac{1}{\sqrt{10.0}}$ and $\theta = 71.6°$ ◊

---

53.   In 1965, Arno Penzias and Robert Wilson discovered that cosmic microwave radiation was left over from the Big Bang expansion of the Universe. Suppose the energy density of this background radiation is 4.00×10^{-14} J/m^3. Determine the corresponding electric field amplitude.

Solution $u = \dfrac{1}{2}\epsilon_0 E_{max}^{\ 2}$ (Eq. 24.27)

$$E_{max} = \sqrt{\frac{2u}{\epsilon_0}} = \sqrt{\frac{2\left(4.00 \times 10^{-14}\ \text{N/m}^2\right)}{8.85 \times 10^{-12}\ \text{C}^2/\text{N}\cdot\text{m}^2}} = 95.1\ \text{mV/m}$$ ◊

---

55.   A linearly polarized microwave of wavelength 1.50 cm is directed along the positive x axis. The electric field vector has a maximum value of 175 V/m and vibrates in the xy plane. (a) Assume that the magnetic field component of the wave can be written in the form $B = B_{max} \sin(kx - \omega t)$ and give values for B_{max}, k, and ω. Also, determine in which plane the magnetic field vector vibrates. (b) Calculate the average value of the Poynting vector for this wave. (c) What radiation pressure would this wave exert if it were directed at normal incidence onto a perfectly reflecting sheet? (d) What acceleration would be imparted to a 500-g sheet (perfectly reflecting and at normal incidence) with dimensions of 1.00 m × 0.750 m?

Solution

(a) $B_{max} = \dfrac{E_{max}}{c} = \dfrac{175\ \text{V/m}}{3.00\times10^8\ \text{m/s}} = 5.83\times10^{-7}\ \text{T}$ ◊

The magnetic field is in the z direction so that $\mathbf{S}=(1/\mu_0)\mathbf{E}\times\mathbf{B}$ can be in the $\mathbf{j}\times\mathbf{k}=\mathbf{i}$ direction.

$k = \dfrac{2\pi}{\lambda} = \dfrac{2\pi}{0.015\ \text{m}} = 419\ \text{m}^{-1}$ ◊

$\omega = kc = \left(419\ \text{m}^{-1}\right)\left(3.00\times10^8\ \text{m/s}\right) = 1.26\times10^{11}\ \text{rad/s}$ ◊

(b) $S_{av} = \dfrac{E_{max}B_{max}}{2\mu_0} = \dfrac{(175\ \text{V/m})\left(5.83\times10^{-7}\ \text{T}\right)}{2\left(4\pi\times10^{-7}\ \text{N/A}^2\right)} = 40.6\ \text{W/m}^2$ ◊

(c) For perfect reflection,

$P_r = \dfrac{2S}{c} = \dfrac{2\left(40.6\ \text{W/m}^2\right)}{3.00\times10^8\ \text{m/s}} = 2.71\times10^{-7}\ \text{N/m}^2$ ◊

(d) We use the particle under a net force model.

$a = \dfrac{F}{m} = \dfrac{P_r A}{m} = \dfrac{\left(2.71\times10^{-7}\ \text{N/m}^2\right)\left(0.750\ \text{m}^2\right)}{0.500\ \text{kg}} = 4.06\times10^{-7}\ \text{m/s}^2$ ◊

57. An astronaut, stranded in space 10.0 m from his spacecraft and at rest relative to it, has a mass (including equipment) of 110 kg. Because he has a 100-W light source that forms a directed beam, he considers using the beam as a photon rocket to propel himself continuously toward the spacecraft. (a) Calculate how long it takes him to reach the spacecraft by this method. (b) Suppose, instead, that he decides to throw the light source away in a direction opposite the spacecraft. If the mass of the light source is 3.00 kg and, after being thrown, moves at 12.0 m/s **relative to the recoiling astronaut**, how long does it take for the astronaut to reach the spacecraft?

Solution   Based on our everyday experience, the force exerted by photons is too small to feel, so it may take a very long time (maybe days!) for the astronaut to travel 10 m with his "photon rocket." Using the momentum of the thrown light seems like a better solution, but it will still take a while (maybe a few minutes) for the astronaut to reach the spacecraft because his mass is so much larger than the mass of the light source.

In part (a), the radiation pressure can be used to find the force that accelerates the astronaut toward the spacecraft. In part (b), the principle of conservation of momentum can be applied to find the time required to travel the 10 m.

(a)   Light exerts on the astronaut a pressure

$$P = F / A = S / c$$

and a force $F = \dfrac{SA}{c} = \dfrac{\mathcal{P}}{c}$:

$$F = \dfrac{100 \text{ J/s}}{3.00 \times 10^8 \text{ m/s}} = 3.33 \times 10^{-7} \text{ N}$$

We use the particle under a net force model.

By Newton's 2nd law, $a = F / m$

$$a = \dfrac{3.33 \times 10^{-7} \text{ N}}{110 \text{ kg}} = 3.03 \times 10^{-9} \text{ m/s}^2$$

We use the particle under constant acceleration model.   The distance traveled is $\Delta x = \dfrac{1}{2} a \Delta t^2$,

and the amount of time it travels is

$$\Delta t = \sqrt{\dfrac{2\Delta x}{a}} = \sqrt{\dfrac{2(10.0 \text{ m})}{3.03 \times 10^{-9} \text{ m/s}^2}}$$

$$\Delta t = 8.12 \times 10^4 \text{ s} = 22.6 \text{ h} \qquad \lozenge$$

(b)   We use the momentum version of the isolated system model, applied to the astronaut and light source as a system. There are no external forces, so the astronaut's momentum before throwing the light is the same as the system momentum afterwards when the now 107-kg astronaut is moving at speed v towards the spacecraft and the light is moving away from the spacecraft at $(12.0 \text{ m/s} - v)$. Thus, $\mathbf{p}_i = \mathbf{p}_f$ gives

$$0 = (107 \text{ kg})v - (3.00 \text{ kg})(12.0 \text{ m/s} - v) \qquad v = \dfrac{36.0 \text{ kg} \cdot \text{m/s}}{110 \text{ kg}} = 0.327 \text{ m/s}$$

We use the particle under constant velocity model $\Delta t = \dfrac{\Delta x}{v} = \dfrac{10.0 \text{ m}}{0.327 \text{ m/s}} = 30.6 \text{ s} \qquad \lozenge$

Related Question   Throwing the light away is certainly a more expedient way to reach the spacecraft, but there is not much chance of retrieving the lamp unless it has a very long cord. How long would the cord need to be, and does its length depend on how hard the astronaut throws the lamp? (You should verify that the minimum cord length is 367 m, independent of the speed that the lamp is thrown.)

---

59. In the absence of cable input or a satellite dish, a television set can use a dipole-receiving antenna for VHF channels and a loop antenna for UHF channels. The UHF antenna produces an emf from the changing magnetic flux through the loop. The TV station broadcasts a signal with a frequency f, and the signal has an electric field amplitude E_{max} and a magnetic field amplitude B_{max} at the location of the receiving antenna. (a) Using Faraday's law, derive an expression for the amplitude of the emf that appears in a single-turn, circular loop antenna with a radius r, which is small compared with the wavelength of the wave. (b) If the electric field in the signal points vertically, what orientation of the loop produces the best reception?

Solution

We can approximate the magnetic field as uniform over the area of the loop while it oscillates in time as $B = B_{max} \cos \omega t$. The induced voltage is

$$\mathcal{E} = -\frac{d\Phi_B}{dt} = -\frac{d}{dt}(BA\cos\theta) = -A\frac{d}{dt}(B_{max}\cos\omega t \cos\theta)$$

$$\mathcal{E} = AB_{max}\omega(\sin\omega t \cos\theta)$$

$$\mathcal{E}(t) = 2\pi f B_{max} A\sin 2\pi f t \cos\theta = 2\pi^2 r^2 f B_{max} \cos\theta \, \sin 2\pi f t$$

(a)   The amplitude of this emf is $\mathcal{E}_{max} = 2\pi^2 r^2 f B_{max} \cos\theta$,
     where θ is the angle between the magnetic field and the normal to the loop.   ◊

(b)   If **E** is vertical, then **B** is horizontal, so the plane of the loop should be vertical, and the plane should contain the line of sight to the transmitter. This will make $\theta = 0°$, so $\cos\theta$ takes on its maximum value.   ◊

Chapter 25

Reflection and Refraction of Light

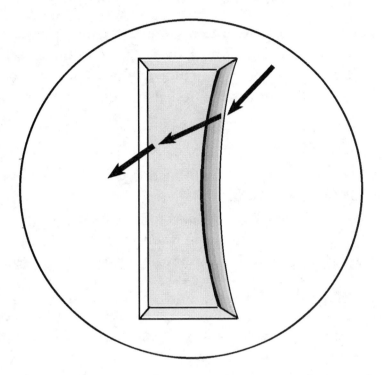

INTRODUCTION

Our study of electromagnetism in Chapter 24 established our understanding of electromagnetic radiation and provides a connection between electromagnetism and optics. We begin our study of optics by examining the behavior of light at the interface between two media; and in the next chapter study the formation of images by flat and curved surfaces through the phenomena of refraction and reflection.

NOTES FROM SELECTED CHAPTER SECTIONS

25.2 The Ray Model in Geometric Optics

In the simplification model called the **ray model** or ray approximation, it is assumed that a wave travels through a medium along a straight line in the direction of its rays. A ray for a given wave is a straight line perpendicular to the wavefront; geometric models are based on these straight lines. This approximation also neglects diffraction effects.

25.3 The Wave Under Reflection

Specular reflection occurs when the reflecting surface is smooth; reflection from a rough surface is called **diffuse reflection**.

In the case of reflection from a smooth surface, the angle of incidence equals the angle of reflection; the angles are measured between the normal to the surface and the respective rays. **The path of a light ray is reversible**, an important property in geometric optics.

25.4 The Wave Under Refraction

Refraction refers to the change in direction of a light ray upon striking obliquely the interface between two transparent media. As a ray travels from one medium to another, the speed of the wave changes but the **frequency does not change**. The angle of refraction (between the refracted ray and the normal) depends on the properties of the two media and the angle of incidence as expressed by Snell's law, Eq. 25.7. The incident ray, reflected ray, and refracted (transmitted) ray lie in the same plane.

25.5 Dispersion and Prisms

For a given material, the index of refraction is a function of the wavelength of the light passing through the material. (The wavelength depends in turn on the speed.) This effect is called dispersion. In particular, when light passes through a prism, a given ray is refracted at two surfaces and emerges bent away from its original direction by an angle of deviation, δ. Due to dispersion, δ is different for different wavelengths.

25.6 Huygens's Principle

Every point on a given wavefront can be considered as a point source for a **secondary wavelet**. At some later time, the new position of the wavefront is determined by the surface tangent to the set of secondary wavelets.

25.7 Total Internal Reflection

Total internal reflection, illustrated in Figure 25.1, is possible only when light rays traveling in one medium are incident on a boundary between the first medium and a second medium of **lesser** index of refraction. The angle θ_c shown in the figure is called the **critical angle**.

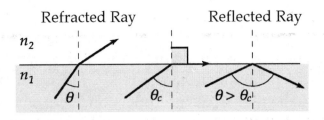

Figure 25.1 Total internal reflection of light occurs at angles of incidence $\theta_1 \geq \theta_c$ where $n_1 > n_2$

EQUATIONS AND CONCEPTS

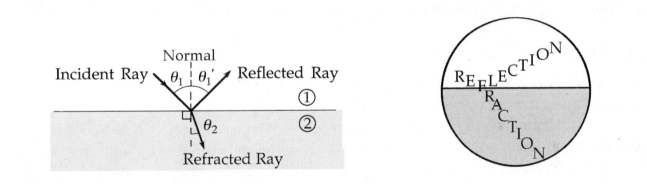

Figure 25.2 The reflection and refraction of light waves at an interface. The directions of the waves are perpendicular to the wavefronts.

Consider the situation in Figure 25.2, in which a light ray is incident obliquely on a smooth, planar surface that forms the boundary between two transparent media of different indices of refraction. A portion of the ray will be reflected back into the original medium, while the remaining fraction will be transmitted into the second medium. The incident, reflected, and refracted rays and the normal to the interface are all in the same plane.

The law of reflection states that the angle of incidence (the angle measured between the incident ray and the normal to the surface) equals the angle of reflection (the angle measured between the reflected ray and the normal to the surface).

$$\theta_1' = \theta_1 \tag{25.1}$$

This is one form of the statement of Snell's law. The angle of refraction (measured relative to the normal to the surface) depends on the angle of incidence, and also on the ratio of the speeds of light in the two media on either side of the refracting surface.

$$\frac{\sin\theta_2}{\sin\theta_1} = \frac{v_2}{v_1} = \text{constant} \tag{25.2}$$

This is the most widely used and most practical form of Snell's law. This equation involves a parameter called the index of refraction, n, the value of which is characteristic of a particular medium. The index of refraction is defined as the ratio of the speed of light in a vacuum to the average speed of light in the medium.

$$n_1 \sin\theta_1 = n_2 \sin\theta_2 \tag{25.7}$$

The frequency of a wave is characteristic of the source. Therefore, as light travels from one medium into another of different index of refraction, the frequency remains constant but the wavelength changes. The index of refraction of a given medium can be expressed as the ratio of the wavelength of light in vacuum to the wavelength in that medium.

$$n = \frac{\lambda_0}{\lambda_n} \tag{25.6}$$

For angles of incidence equal to or greater than the critical angle, the incident ray will be totally internally reflected back into the first medium. Total internal reflection is possible only when a light ray is directed from a medium of high index of refraction into a medium of lower index of refraction.

$$\sin\theta_c = \frac{n_2}{n_1} \text{ (for } n_1 > n_2\text{)} \qquad (25.8)$$

SUGGESTIONS, SKILLS, AND STRATEGIES

It is helpful in this chapter to review the laws of trigonometry. Most important, of course, is the definition of the trigonometric functions:

$$\tan\theta = \frac{\text{Opposite length}}{\text{Adjacent length}}$$

$$\sin\theta = \frac{\text{Opposite length}}{\text{Hypotenuse}} \qquad \sin(-\theta) = -\sin\theta$$

$$\cos\theta = \frac{\text{Adjacent length}}{\text{Hypotenuse}} \qquad \cos(-\theta) = \cos\theta$$

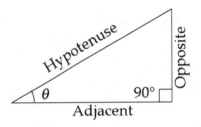

(Valid for right triangles only)

It will be useful to remember that the interior angles of a triangle add to 180°. It may also be useful to convert trigonometric functions from one form to another, using the trigonometric versions of the Pythagorean theorem:

$$\sin^2\theta = 1 - \cos^2\theta \qquad\qquad \tan^2\theta = \frac{\sin^2\theta}{1 - \sin^2\theta}$$

Finally, you should review the angle-sum relations:

$$\sin(\theta + \phi) = \sin\theta\cos\phi + \cos\theta\sin\phi \qquad\qquad \cos(\theta + \phi) = \cos\theta\cos\phi - \sin\theta\sin\phi$$

REVIEW CHECKLIST

▷ Understand Huygens's principle and the use of this technique to construct the subsequent position and shape of a given wavefront.

▷ Determine the directions of the reflected and refracted rays when a light ray is incident obliquely on the interface between two optical media.

▷ Understand the conditions under which total internal reflection can occur in a medium and determine the critical angle for a given pair of adjacent media.

ANSWERS TO SELECTED CONCEPTUAL QUESTIONS

4. As light travels from a vacuum ($n = 1$) to a medium such as glass ($n > 1$), does the wavelength of the light change? Does the frequency change? Does the speed change? Explain.

Answer

You can think about this based upon the fact that the light wave must be continuous. Therefore, the frequency of the light must remain constant, since any one crest coming up to the interface must create one crest in the new medium. However, the speed of light does vary in different media, as described by Snell's law. In addition, since the speed of the light wave is equal to the product of frequency and wavelength, $v = f\lambda$, and the frequency does not change, we can be sure that the wavelength must change. The speed decreases upon entering the new medium; therefore the wavelength decreases as well.

6. Explain why a diamond loses most of its sparkle when submerged in carbon disulfide and why an imitation diamond of cubic zirconia loses all of its sparkle in corn syrup.

Answer

A ball covered with mirrors sparkles by reflecting light from its surface. On the other hand, a faceted diamond lets in light at the top, reflects it by total internal reflection in the bottom half, and sends the light out through the top again. Its high index of refraction means that in air, the critical angle for total internal reflection, $\theta_c = \sin^{-1}(n_{air}/n_{diamond})$, is small. Thus, light rays enter through a large area, and exit through a very small area with a much higher intensity. When a diamond is immersed in carbon disulfide, the critical angle is increased to $\theta_c = \sin^{-1}(n_{CS_2}/n_{diamond})$. As a result, the light is emitted from the diamond over a larger area, and appears less intense.

An imitation diamond of cubic zirconia loses all of its sparkle in corn syrup because its index of refraction is the same as that of corn syrup. This stone will therefore reflect almost no light, either internally or externally. If the absorption spectrum (color) of the stone is the same as that of the corn syrup, we can expect the stone to be invisible.

12. Explain why a diamond sparkles more than a glass crystal of the same shape and size.

Answer

Diamond has a larger index of refraction than glass, and consequently has a smaller critical angle for internal reflection. This results in a greater amount of light being internally reflected, and more light being emitted within the critical angles.

☐ ☐ ☐ ☐

18. Suppose you are told that only two colors of light (X and Y) are sent through a glass prism and that X is bent more than Y. Which color travels more slowly in the prism?

Answer

If the light slows down upon entering the prism, a light ray that is bent more suffers a greater loss of speed as it enters the new medium. Therefore, the X light rays travel more slowly.

☐ ☐ ☐ ☐

SOLUTIONS TO SELECTED END-OF-CHAPTER PROBLEMS

3. An underwater scuba diver sees the Sun at an apparent angle of 45.0° from the vertical. What is the actual direction of the Sun?

Solution We use the wave under refraction model. Refraction happens as sunlight in air crosses into water. The interface is horizontal, so the normal is vertical:

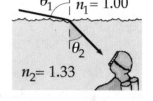

$$n_1 \sin \theta_1 = n_2 \sin \theta_2$$

gives $\qquad \sin \theta_1 = 1.333 \sin 45.0° \qquad\qquad \sin \theta_1 = (1.333)(0.707) = 0.943$

The sunlight is at $\theta_1 = 70.5°$ to the vertical, so the Sun is 19.5° above the horizon. ◊

---

7. A ray of light strikes a flat block of glass ($n = 1.50$) of thickness 2.00 cm at an angle of 30.0° with the normal. Trace the light beam through the glass, and find the angles of incidence and refraction at each surface.

Solution At entry: $\qquad n_1 \sin \theta_1 = n_2 \sin \theta_2 \qquad\qquad 1.00 \sin 30° = 1.50 \sin \theta_2$

and $\qquad\qquad \theta_2 = \sin^{-1}\left(\dfrac{0.500}{1.50}\right) = 19.5°$ ◊

To do geometric optics, you must remember some geometry. The surfaces of entry and exit are parallel so their normals are parallel. Then angle θ_2 of refraction at entry and the angle θ_3 of incidence at exit are alternate interior angles formed by the ray as a transversal cutting parallel lines.

So $\qquad\qquad \theta_3 = \theta_2 = 19.5°$ ◊

At the exit, $\qquad\qquad n_2 \sin \theta_3 = n_1 \sin \theta_4 \qquad 1.5 \sin 19.5° = 1 \sin \theta_4$

and $\qquad\qquad \theta_4 = 30.0°$ ◊

The exiting ray in air is parallel to the original ray in air. Thus, a car windshield of uniform thickness will not distort, but shows the driver the actual direction to every object outside.

---

21. The refractive index for violet light in silica flint glass is 1.66, and that for red light is 1.62. A prism has an apex angle of 60.0°, measured between the surface at which the light enters the prism and the surface at which the light leaves the prism. What is the angular dispersion of visible light passing through the prism if the angle of incidence is 50.0° (See Fig. P25.21)?

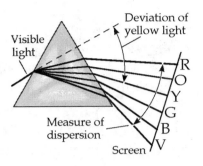

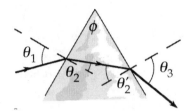

Figure P25.21

Solution

Call the angles of incidence and refraction, at the surfaces of entry and exit, θ_1, θ_2, θ_2', and θ_3, in the order as shown. The apex angle ($\phi = 60.0°$) is the angle between the surfaces of entry and exit. The ray in the glass forms a triangle with these surfaces, in which the interior angles must add to 180°.

Thus,
$$\left(90° - \theta_2\right) + \phi + \left(90 - \theta_2'\right) = 180°$$

and
$$\theta_2' = \phi - \theta_2$$

This is a general rule for light going through prisms.

For the incoming ray, $\quad \sin\theta_2 = \dfrac{\sin\theta_1}{n}: \qquad \left(\theta_2\right)_{\text{violet}} = \sin^{-1}\left(\dfrac{\sin 50.0°}{1.66}\right) = 27.48°$

$$\left(\theta_2\right)_{\text{red}} = \sin^{-1}\left(\dfrac{\sin 50.0°}{1.62}\right) = 28.22°$$

For the outgoing ray, $\quad \theta_2' = 60° - \theta_2 \quad$ and $\quad \sin\theta_3 = n\sin\theta_2'$

$$\left(\theta_3\right)_{\text{violet}} = \sin^{-1}\left[1.66 \sin 32.52°\right] = 63.17°$$

$$\left(\theta_3\right)_{\text{red}} = \sin^{-1}\left[1.62 \sin 31.78°\right] = 58.56°$$

The dispersion is the difference between these two angles:

$$\Delta\theta_3 = 63.17° - 58.56° = 4.61° \qquad \Diamond$$

22. A triangular glass prism with an apex angle $\Phi = 60.0°$ has an index of refraction $n = 1.50$ (Fig. P25.22). What is the smallest angle of incidence θ_1 for which a light ray can emerge from the other side?

23. A triangular glass prism with apex angle Φ has index of refraction n (see Fig. P25.22). What is the smallest angle of incidence θ_1 for which a light ray can emerge from the other side?

Solution

Call the angles of incidence and refraction, at the surfaces of entry and exit, θ_1, θ_2, θ_2', and θ_3, in order as shown. The apex angle ϕ is the angle between the surfaces of entry and exit. The ray in the glass forms a triangle with these surfaces, in which the interior angles must add to 180°. Thus, with

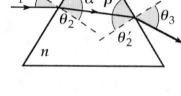

$$\phi = 60.0°,$$

$$(90 - \theta_2) + 60 + (90 - \theta_2') = 180$$

so $\theta_2 + \theta_2' = 60.0°$ (1)

which is a general rule for light going through prisms. At the first refraction, Snell's law gives

$$\sin\theta_1 = 1.50 \sin\theta_2$$ (2)

At the second boundary, we want total internal reflection:

$$1.50 \sin\theta_2' = 1.00 \sin 90° = 1.00$$

or $\theta_2' = \sin^{-1}(1.00/1.50) = 41.8°$

Now by equation (1) above,

$$\theta_2 = 60.0° - 41.8° = 18.2°$$

$$\phi = \Phi$$

$$(90 - \theta_2) + \Phi + (90 - \theta_2') = 180$$

so $\theta_2 + \theta_2' = \Phi$ (1)

which is a general rule for light going through prisms. At the first interface between the air and prism,

$$\sin\theta_1 = n \sin\theta_2$$ (2)

At the second interface, we want total internal reflection:

$$n \sin\theta_2' = \sin 90° = 1$$

or $\theta_2' = \sin^{-1}(1/n)$

Now by equation (1) above,

$$\theta_2 = \Phi - \theta_2' = \Phi - \sin^{-1}(1/n)$$

while by equation (2), we find that

$$\theta_1 = \sin^{-1}(1.50 \sin 18.2°)$$

So $\quad \theta_1 = 27.9°$ ◊

Figure 25.24 in the text shows the effect nicely.

Note that in the case of this problem, the numeric solution eliminated several of the analytic steps of the variable solution, and thus was much shorter. Whether a given problem is best solved analytically or numerically depends on the complexity of the problem.

while by equation (2), we find that

$$\theta_1 = \sin^{-1}\left(n \sin\left(\Phi - \sin^{-1}(1/n)\right)\right)$$

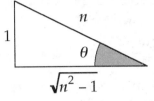

Looking at the drawing of a right triangle, remember the identities that

$$\sin(\alpha - \beta) = \sin\alpha \ \cos\beta \ - \cos\ \alpha \ \sin\beta$$

$$\cos\left(\sin^{-1}(1/n)\right) = \sqrt{n^2 - 1}/n$$

θ_1 therefore simplifies to

$$\theta_1 = \sin^{-1}\left(\left(\sqrt{n^2 - 1}\right)\sin\Phi - \cos\Phi\right) \quad ◊$$

---

27. Consider a common mirage formed by super-heated air just above a roadway. A truck driver whose eyes are 2.00 m above the road, where $n = 1.000\ 3$, looks forward. She perceives the illusion of a patch of water ahead on the road, where her line of sight makes an angle of 1.20° below the horizontal. Find the index of refraction of the air just above the road surface. (**Hint:** Treat this as a problem in total internal reflection.)

Solution Think of the air as in two discrete layers, the first medium being cooler air with $n_1 = 1.0003$ and the second medium being hot air with a lower index, which reflects light from the sky by total internal reflection. Use:

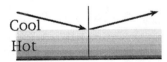

$$n_1 \sin\theta_1 \geq n_2 \sin 90° \qquad\qquad 1.0003 \sin 88.8° \geq n_2$$

$$n_2 \leq 1.00008 \qquad\qquad\qquad\qquad\qquad ◊$$

---

31. A laser beam strikes one end of a slab of material, as shown in Figure P25.31. The index of refraction of the slab is 1.48. Determine the number of internal reflections of the beam before it emerges from the opposite end of the slab.

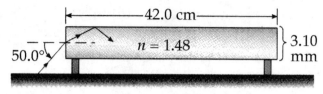

Figure P25.31

Solution   Upon entry,

$$\sin 50.0° = 1.48 \sin\theta_2$$

so

$$\theta_2 = 31.17°$$

The beam strikes the top face at horizontal coordinate

$$x_1 = \frac{1.55 \text{ mm}}{\tan 31.17°} = 2.562 \text{ mm}$$

According to the wave under reflection model, the beam makes the same angle with the horizontal after every reflection.

After its first reflection, the beam strikes a face every

$$2x_1 = 5.124 \text{ mm}$$

Since the slab is 420 mm long, the beam makes another

$$\frac{420 - 2.562}{5.124} = 81 \text{ reflections}$$

Therefore, the total is

82 reflections   ◊

---

35. A small underwater pool light is 1.00 m below the surface. The light emerging from the water forms a circle on the water's surface. What is the diameter of this circle?

Solution   At the edge of the circle, the light is totally internally reflected:

$$n_1 \sin\theta_c = n_2 \sin 90°$$

$$1.333 \sin\theta_c = 1$$

$$\theta_c = \sin^{-1}(0.750) = 48.6°$$

The radius then satisfies $\tan\theta_c = \dfrac{r}{1.00 \text{ m}}$

So the diameter is $d = 2 \tan\theta_c = 2 \tan 48.6° = 2.27 \text{ m}$ ◊

39. A hiker stands on an isolated mountain peak near sunset and observes a rainbow caused by water droplets in the air 8.00 km away. The valley is 2.00 km below the mountain peak and entirely flat. What fraction of the complete circular arc of the rainbow is visible to the hiker? (See Figure 25.17)

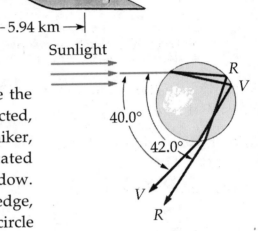

Figure 25.17

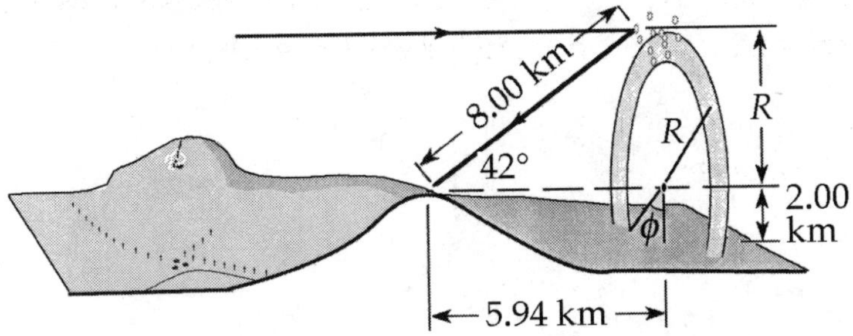

Solution

Horizontal light rays from the setting Sun pass above the hiker. The light rays are twice refracted and once reflected, as in Figure 25.16. The most intense light reaching the hiker, that which represents the visible rainbow, is located between angles of 40.0° and 42.0° from the hiker's shadow. The hiker sees a greater percentage of the violet inner edge, so we consider the red outer edge. The radius R of the circle of droplets is

Figure 25.16

$$R = (8.00 \text{ km}) \sin 42.0° = 5.35 \text{ km}$$

Then the angle ϕ, between the vertical and the radius where the bow touches the ground, is given by

$$\cos\phi = \frac{2.00 \text{ km}}{R} = \frac{2.00 \text{ km}}{5.35 \text{ km}} = 0.374 \qquad \text{or} \qquad \phi = 68.1°$$

The angle filled by the visible bow is $360° - 2(68.1°) = 224°$, so the visible bow is

$$\frac{224°}{360°} = 62.2\% \text{ of a circle} \qquad\qquad \Diamond$$

This striking view motivated Charles Wilson's 1906 invention of the cloud chamber, a standard tool of nuclear physics. Look for a full circle of color around your shadow when you fly in an airplane.

45. The light beam shown in Figure P25.45 strikes surface 2 at the critical angle. Determine the angle of incidence θ_1.

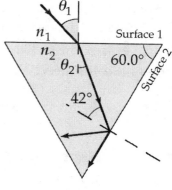

Surface 1

Solution Call the index of refraction of the prism material n_2. Use n_1 to represent the surrounding medium.

At surface 2, $\qquad\qquad n_2 \sin 42° = n_1 \sin 90°$

So $\qquad\qquad \dfrac{n_2}{n_1} = \dfrac{1}{\sin 42°} = 1.49$

Figure P25.45

Call the angle of refraction θ_2 at the surface 1. The ray inside the prism forms with surfaces 1 and 2 a triangle whose interior angles add to 180°. Thus,

$$(90° - \theta_2) + 60° + (90° - 42°) = 180° \qquad \text{and} \qquad \theta_2 = 18.0°$$

At surface 1, $\qquad\qquad n_1 \sin\theta_1 = n_2 \sin 18.0°$

$$\sin\theta_1 = \left(\dfrac{n_2}{n_1}\right)\sin 18.0°$$

$$\sin\theta_1 = 1.49 \sin 18.0°$$

$$\theta_1 = 27.5° \qquad\qquad\qquad\qquad\qquad\qquad \Diamond$$

---

47. A light ray of wavelength 589 nm is incident at an angle θ on the top surface of a block of polystyrene, as shown in Figure P25.47. (a) Find the maximum value of θ for which the refracted ray undergoes total internal reflection at the left vertical face of the block. Repeat the calculation for the case in which the polystyrene block is immersed in (b) water and (c) carbon disulfide.

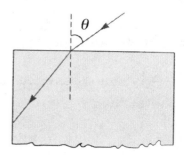

Figure P25.47

Solution

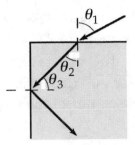

(a) The index of refraction (for 589 nm light) of each material is listed below:

Air:	1.00
Water:	1.33
Polystyrene:	1.49
Carbon disulfide:	1.63

For polystyrene **surrounded by air,**

internal reflection requires
$$\theta_3 = \sin^{-1}\left(\frac{1}{1.49}\right) = 42.2°$$

and then from the geometry,
$$\theta_2 = 90.0° - \theta_3 = 47.8°$$

From Snell's law,
$$\sin\theta_1 = 1.49 \sin 47.8°$$

This has no solution; thus, the real maximum value for θ_1 is 90.0°: total internal reflection always occurs.   ◊

(b) For polystyrene **surrounded by water,** we have
$$\theta_3 = \sin^{-1}\left(\frac{1.33}{1.49}\right) = 63.2°$$

and
$$\theta_2 = 26.8°$$

From Snell's law, $n_1 \sin\theta_1 = n_2 \sin\theta_2$:
$$1.33 \sin\theta_1 = 1.49 \sin 26.8°$$

and
$$\theta_1 = 30.3°$$ ◊

(c) **This is not possible** since the beam is initially traveling in a medium of lower index of refraction.

49. A shallow glass dish is 4.00 cm wide at the bottom, as shown in Figure P25.49. When an observer's eye is placed as shown, the observer sees the edge of the bottom of the empty dish. When this dish is filled with water, the observer sees the center of the bottom of the glass. Find the height of the glass.

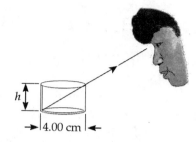

Figure P25.49

Solution

$$\tan\theta_1 = \frac{4.00 \text{ cm}}{h} \qquad \text{and} \qquad \tan\theta_2 = \frac{2.00 \text{ cm}}{h}$$

$$\tan^2\theta_1 = \left(2.00\,\tan\theta_2\right)^2 = 4.00\,\tan^2\theta_2$$

$$\frac{\sin^2\theta_1}{1-\sin^2\theta_1} = 4.00\left(\frac{\sin^2\theta_2}{1-\sin^2\theta_2}\right) \tag{1}$$

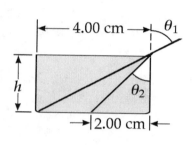

Snell's law in this case is:

$$n_1\sin\theta_1 = n_2\sin\theta_2 \qquad\qquad \sin\theta_1 = 1.333\,\sin\theta_2$$

Squaring both sides,

$$\sin^2\theta_1 = 1.777\,\sin^2\theta_2 \tag{2}$$

Substituting (2) into (1) yields

$$\frac{1.777\,\sin^2\theta_2}{1-1.777\,\sin^2\theta_2} = 4.00\left(\frac{\sin^2\theta_2}{1-\sin^2\theta_2}\right)$$

Defining $x = \sin^2\theta_2$ and solving for x,

$$\frac{0.444}{1-1.777\,x} = \frac{1}{1-x}$$

$$0.444 - 0.444\,x = 1 - 1.777\,x \qquad\qquad x = 0.417$$

From x we can solve for θ_2:

$$\theta_2 = \sin^{-1}\sqrt{0.417} = 40.2°$$

and

$$h = \frac{2.00 \text{ cm}}{\tan\theta_2} = \frac{2.00 \text{ cm}}{\tan 40.2°} = 2.37 \text{ cm} \qquad \lozenge$$

Chapter 26

Image Formation by Mirrors and Lenses

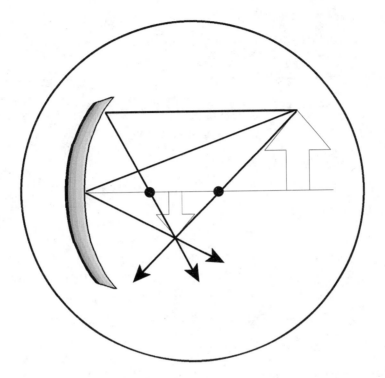

INTRODUCTION

This chapter is concerned with the images formed when light interacts with flat and curved surfaces. Mirrors and lenses form images by the phenomena of reflection and refraction. We continue to use the ray approximation introduced in the preceding chapter and make use of geometric models developed from the principles of reflection and refraction. Mathematical representations will be developed based on these models. Examples of both everyday and scientific devices which form images by reflection and refraction will be described.

NOTES FROM SELECTED CHAPTER SECTIONS

26.1   Images Formed by Flat Mirrors

An image formed by a flat mirror has the following properties:

- The image is as far behind the mirror as the object is in front.

- The image is unmagnified, virtual, and upright. (By upright, we mean that, if the object arrow points upward, so does the image arrow.)

- The image has an apparent right-left reversal.

26.2   Images Formed by Spherical Mirrors

A spherical mirror is a reflecting surface, which has the shape of a segment of a sphere. A concave mirror is one which reflects light from the inner concave surface; a convex mirror reflects light from the outer convex surface. When using the mirror equation, Eq. 26.6, carefully adhere to the sign conventions stated in the **Suggestions, Skills, and Strategies** section.

Real images are formed at a point when **reflected light actually passes through the image point. Virtual images** are formed at a point when light rays **appear to diverge from the image point.**

The point of intersection of any two of the following rays in a ray diagram for mirrors locates the image:

- The first ray is drawn from the top of the object parallel to the principal axis and is reflected back through the focal point, F.

- The second ray is drawn from the top of the object through the focal point and is reflected parallel to the axis.

- The third ray is drawn from the top of the object through the center of curvature, C, and is reflected back on itself.

- The fourth ray is drawn from the top of the object to the center of the mirror's surface and reflects on the other side of the principal axis, with an angle of reflection equal to its angle of incidence.

26.4 Thin Lenses

Images formed by thin lenses can be located and described by pictorial representations called ray diagrams.

To locate the position of an image formed by a converging lens, draw the following three rays from the top of the object:

- Ray 1 is drawn parallel to the principal axis. After being refracted by the lens, this ray passes through the focal point on the back side of the lens.

- Ray 2 is drawn through the center of the lens. This ray continues in a straight line.

- Ray 3 is drawn through the focal point on the front side of the lens (or as if coming from the focal point if $p < f$) and emerges from the lens parallel to the principal axis.

In the case of a diverging lens, the following three rays are drawn from the top of the object:

- Ray 1 is drawn parallel to the principal axis, and after refraction emerges directed away from the focal point on the front side of the lens.

- Ray 2 is drawn through the center of the lens and continues in a straight line.

- Ray 3 is drawn in the direction toward the focal point on the back side of the lens and emerges from the lens parallel to the principle axis.

Note that any two rays can be used too locate the object; the third ray is a useful check on the accuracy of your diagram.

26.5 Lens Aberrations

Aberrations are responsible for the formation of imperfect images by lenses and mirrors. Spherical aberration is due to the variation in focal points for parallel incident rays that strike the lens at various distances from the optical axis. Chromatic aberration arises from the fact that light of different wavelengths focuses at different points when refracted by a lens.

EQUATIONS AND CONCEPTS

In using the equations of this chapter, you must be very careful to use the correct algebraic sign for each quantity. The sign conventions appropriate for the form of the equations stated here are summarized in **SUGGESTIONS, SKILLS AND STRATEGIES**.

The **mirror equation** is used to locate the position of an image formed by reflection of paraxial rays. The focal point of a spherical mirror is located midway between the center of curvature and the center of the mirror.

$$\frac{1}{p} + \frac{1}{q} = \frac{1}{f} \qquad (26.6)$$

$$f = \frac{R}{2} \qquad (26.5)$$

The **lateral magnification** of a spherical mirror can be stated either as a ratio of image size to object size or in terms of the ratio of image distance to object distance.

$$M = \frac{h'}{h} = -\frac{q}{p} \qquad (26.2)$$

A magnified image of an object can be formed by a single spherical refracting surface of radius R, which separates two media whose indices of refraction are n_1 and n_2.

$$\frac{n_1}{p} + \frac{n_2}{q} = \frac{n_2 - n_1}{R} \qquad (26.8)$$

$$M = -\frac{n_1 q}{n_2 p} \qquad (26.9)$$

A special case is that of the virtual image formed by a **planar refracting surface** $(R = \infty)$.

$$\frac{n_1}{p} = -\frac{n_2}{q}$$

$$q = -\frac{n_2}{n_1}p \qquad (26.10)$$

The focal length of a **thin lens** is determined by the characteristic properties of the lens (index of refraction n and radii of curvature R_1 and R_2)..

$$\frac{1}{f} = (n-1)\left(\frac{1}{R_1} - \frac{1}{R_2}\right) \qquad (26.13)$$

The lateral magnification, M, will be negative when the image is inverted.

$$M = \frac{h'}{h} = -\frac{q}{p} \qquad (26.11)$$

The thin lens equation is identical to the mirror equation and can be used with both converging (positive f) and diverging (negative f) lenses.

$$\frac{1}{p} + \frac{1}{q} = \frac{1}{f} \qquad (26.12)$$

SUGGESTIONS, SKILLS, AND STRATEGIES

A major portion of this chapter is devoted to the development and presentation of equations, which can be used to determine the location and nature of images formed by various optical components acting either singly or in combination. It is essential that these equations be used with the correct algebraic sign associated with each quantity involved. You must understand clearly the sign conventions for mirrors, refracting surfaces, and lenses. The following discussion represents a review of these sign conventions.

SIGN CONVENTIONS FOR MIRRORS

Equations:
$$\frac{1}{p} + \frac{1}{q} = \frac{1}{f} = \frac{2}{R} \qquad\qquad M = \frac{h'}{h} = -\frac{q}{p}$$

The front side of the mirror is the region on which light rays are incident and reflected.

---

p is $+$ if the object is in front of the mirror (real object).
p is $-$ if the object is in back of the mirror (virtual object).

q is $+$ if the image is in front of the mirror (real image).
q is $-$ if the image is in back of the mirror (virtual image).

Both f and R are $+$ if the center of curvature is in front of the mirror (concave mirror).
Both f and R are $-$ if the center of curvature is in back of the mirror (convex mirror).

If M is positive, the image is upright.
If M is negative, the image is inverted.

You should check the sign conventions as stated against the situations described in Figure 26.1. You may also wish to check the image location in each diagram by drawing in ray number two, as described in Section 26.2.

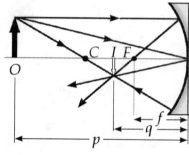

(a) Concave Mirror

+: $p, q, f,$ and R

Image real
  and inverted

(b) Concave Mirror

+: p, f, R
−: q

Image virtual, upright,
  and enlarged

(c) Convex Mirror

+: p
−: $q, f,$ and R

Image virtual, upright,
  and diminished

Figure 26.1   Figures describing sign conventions for mirrors.

SIGN CONVENTIONS FOR REFRACTING SURFACES

Equations:
$$\frac{n_1}{p} + \frac{n_2}{q} = \frac{n_2 - n_1}{R} \qquad\qquad M = \frac{h'}{h} = -\frac{n_1 q}{n_2 p}$$

In the following table, the **front** side of the surface is the side **from which the light is incident**.

---

p is + if the object is in front of the surface (real object).
p is – if the object is in back of the surface (virtual object).

q is + if the image is in back of the surface (real image).
q is – if the image is in front of the surface (virtual image).

R is + if the center of curvature is in back of the surface.
R is – if the center of curvature is in front of the surface.

n_1 refers to the index of refraction of the medium on the side of the interface from which the light comes.

n_2 is the index of refraction of the medium into which the light is transmitted after refraction at the interface.

Review the above sign conventions for the situations shown in Figure 26.2.

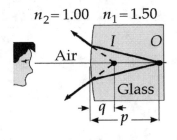

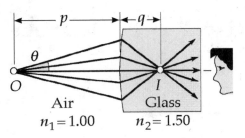

$p +$ (real object)
$q -$ (virtual image)
$R -$ (concave to incident light)
n_1 and n_2 as shown

$p +$ (real object)
$q +$ (real image)
$R +$ (convex to incident light)
n_1 and n_2 as shown

Figure 26.2 These figures describe sign conventions for refracting surfaces. In the first case, the object appears closer than it actually is. In the second case, an object beyond the glass appears to be in the glass.

SIGN CONVENTIONS FOR THIN LENSES

Equations:
$$\frac{1}{p} + \frac{1}{q} = \frac{1}{f} = (n-1)\left(\frac{1}{R_1} - \frac{1}{R_2}\right) \qquad M = \frac{h'}{h} = -\frac{q}{p}$$

In the following table, the **front** of the lens is the **side from which the light is incident**.

---

p is + if the object is in front of the lens.
p is − if the object is in back of the lens.

q is + if the image is in back of the lens.
q is − if the image is in front of the lens.

f is + if the lens is thickest at the center.
f is − if the lens is thickest at the edges.

R_1 and R_2 are + if the center of curvature is in back of the lens.
R_1 and R_2 are − if the center of curvature is in front of the lens.

The sign conventions for thin lenses are illustrated by the examples shown in Figure 26.3.

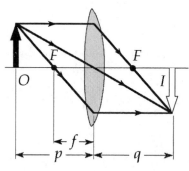

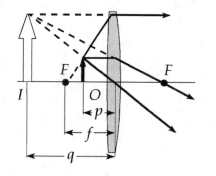

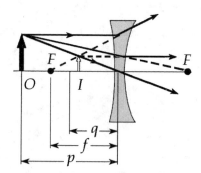

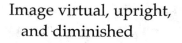

(a) Double-Convex Lens
(Converging)

+: p, q, f, R_1
−: R_2

Image real
  and inverted

(b) Double-Convex Lens
(Converging)

+: $p, R_1, f,$
−: q, R_2

Image virtual, upright,
  and enlarged

(c) Double-Concave Lens
(Diverging)

+: p, R_2
−: q, R_1, f

Image virtual, upright,
  and diminished

Figure 26.3   Figures describing the sign conventions for various thin lenses.

REVIEW CHECKLIST

▷ Identify the following properties which characterize an image formed by a lens or mirror system with respect to an object: position, magnification, orientation (i.e. inverted or upright) and whether real or virtual.

▷ Understand the relationship of the algebraic signs associated with calculated quantities to the nature of the image and object: real or virtual, upright or inverted.

▷ Calculate the location of the image of a specified object as formed by a plane mirror, spherical mirror, plane refracting surface, spherical refracting surface, thin lens, or a combination of two or more of these devices. Determine the magnification and character of the image in each case.

▷ Construct ray diagrams to determine the location and nature of the image of a given object when the geometrical characteristics of the optical device (lens or mirror) are known.

ANSWERS TO SELECTED CONCEPTUAL QUESTIONS

2.   The sideview mirror on late-model cars warns the user that objects may be closer than they appear. What kind of mirror is being used and why was this type selected?

Answer   The mirror would probably be a convex mirror. This type would have been selected because the designers wish to give the driver a wide field of view, and a virtual upright image for all distances.

7.   Explain why a fish in a spherical goldfish bowl appears larger than it really is.

Answer   A spherical goldfish bowl acts as a single refracting surface; the magnification of any object in the bowl can be calculated from Equations 26.8 and 26.9 to be

$$M = 1 + \left(\frac{n_w - 1}{R}\right)q = 1 + \frac{0.33\,q}{R} \qquad \text{where } R \text{ is negative.}$$

Since the image distance q is also always negative, this gives a magnification that is always greater than 1. Therefore, the image of the goldfish will always be larger than the object.

10.   Lenses used in eyeglasses, whether converging or diverging, are always designed such that the middle of the lens curves away from the eye, like the center lenses of Figure 26.18a and 26.18b. Why?

Answer   With the meniscus design, when you direct your gaze near the outer circumference of the lens you receive a ray that has passed through glass with more nearly parallel surfaces of entry and exit. Thus, the lens minimally distorts the direction to the object you are looking at. If you wear glasses, you can demonstrate this by turning them around and looking through them the wrong way, maximizing the distortion.

22.   A classic science fiction story, "The Invisible Man", tells of a person who becomes invisible by changing the index of refraction of his body to that of air. This story has been criticized by students who know how the eye works; they claim the invisible man would be unable to see. On the basis of your knowledge of the eye, could he see or not?

Answer   The cornea of the eye works by refracting light, and creating a real image at the retina. Therefore, unless the invisible man wore a pair of glasses that focused the light, everything he saw would be a blur, and an image would not be formed.

This is actually a very real problem for some former cataract patients. The cornea, which turns from clear to a milky-white in the case of cataracts, can be removed, but the patient must then either have it replaced, or must wear very thick glasses.

SOLUTIONS TO SELECTED END-OF-CHAPTER PROBLEMS

3. Determine the minimum height of a vertical flat mirror in which a person 5'10" in height can see his or her full image. (A ray diagram would be helpful.)

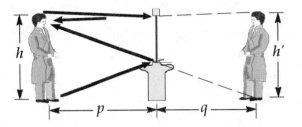

Solution The flatness of the mirror is described by $R = \infty$, $f = \infty$, and $1/f = 0$. By our general mirror equation,

$$\frac{1}{p} + \frac{1}{q} = \frac{1}{f} = 0 \qquad \text{or} \qquad q = -p$$

Thus, the image is as far behind the mirror as the person is in front. The magnification is then

$$M = \frac{-q}{p} = 1 = \frac{h'}{h} \qquad \text{so} \qquad h' = h = 70 \text{ inches}$$

The required height of the mirror is defined by the triangle from the person's eyes to the top and bottom of his image, as shown. From the geometry of the triangle, we see that the mirror height must be:

$$h'\left(\frac{p}{p-q}\right) = h'\left(\frac{p}{2p}\right) = \frac{h'}{2}$$

Thus, the mirror must be at least 35 inches high. ◊

---

9. A concave mirror has a radius of curvature of 60.0 cm. Calculate the image position and magnification of an object placed in front of the mirror (a) at a distance of 90.0 cm and (b) at a distance of 20.0 cm. (c) In each case, draw ray diagrams to obtain the image characteristics.

Solution It is always a good idea to first draw a ray diagram for any optics problem. This gives a qualitative sense of the character and position of the image relative to the object. From the ray diagrams on the next page, we see that when the object is 90 cm from the mirror, the image will be real, inverted, diminished, and about 45 cm in front of the mirror, midway between the center of curvature and the focal point. When the object is 20 cm from the mirror, the image is virtual, upright, magnified, and located about 50 cm behind the mirror.

The mirror equation can be used to find precise quantitative values.

(a) The mirror equation is applied using the sign conventions listed in **Suggestions, Skills, and Strategies**:

$$\frac{1}{p} + \frac{1}{q} = \frac{2}{R} \qquad \text{or} \qquad \frac{1}{90.0 \text{ cm}} + \frac{1}{q} = \frac{2}{60.0 \text{ cm}}$$

$$\frac{1}{q} = \frac{2}{60.0 \text{ cm}} - \frac{1}{90.0 \text{ cm}} = 0.0222 \text{ cm}^{-1}$$

$q = 45.0$ cm (real, in front of the mirror)   ◊

$$M = -\frac{q}{p} = -\frac{45.0 \text{ cm}}{90.0 \text{ cm}} = -0.500 \text{ (inverted)}$$ ◊

(b) We again use the mirror equation:

$$\frac{1}{p} + \frac{1}{q} = \frac{2}{R} \qquad \text{or} \qquad \frac{1}{20.0 \text{ cm}} + \frac{1}{q} = \frac{2}{60.0 \text{ cm}}$$

$$\frac{1}{q} = \frac{2}{60.0 \text{ cm}} - \frac{1}{20.0 \text{ cm}} = -0.0167 \text{ cm}^{-1}$$

$q = -60.0$ cm   (virtual, behind the mirror)   ◊

$$M = -\frac{q}{p} = -\frac{(-60.0 \text{ cm})}{20.0 \text{ cm}} = 3.00 \text{ (upright)}$$ ◊

(c) Each ray diagram is shown next to its solution.   ◊

The calculated image characteristics agree well with our predictions. It is easy to miss a minus sign or to make a computational mistake when using the mirror-lens equations, so the qualitative values obtained from the ray diagrams are useful for a check on the reasonableness of the calculated values.

11. A spherical convex mirror (Figure P26.11) has a radius of curvature with a magnitude of 40.0 cm. Determine the position of the virtual image and magnification (a) for an object distance of 30.0 cm and (b) for an object distance of 60.0 cm. (c) Are the images upright or inverted?

Solution

The convex mirror is described by $f = \dfrac{R}{2} = \dfrac{-40.0 \text{ cm}}{2} = -20.0 \text{ cm}$

(a) $\dfrac{1}{p} + \dfrac{1}{q} = \dfrac{1}{f}$,   so $\dfrac{1}{30.0 \text{ cm}} + \dfrac{1}{q} = \dfrac{1}{-20.0 \text{ cm}}$ and $q = -12.0 \text{ cm}$ ◊

$M = -\dfrac{q}{p} = -\left(\dfrac{-12.0 \text{ cm}}{30.0 \text{ cm}}\right) = +0.400$ ◊

(c)   The image is behind the mirror, upright, virtual, and diminished.   ◊

(b) $\dfrac{1}{60.0 \text{ cm}} + \dfrac{1}{q} = \dfrac{1}{-20.0 \text{ cm}}$

$q = -15.0 \text{ cm}$ ◊

$M = \dfrac{-q}{p} = -\left(\dfrac{-15.0 \text{ cm}}{60.0 \text{ cm}}\right) = +0.250$ ◊

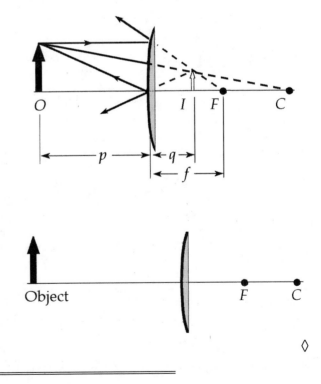

(c)   The principal ray diagram is an essential complement to the numerical description of the image. Add rays on this diagram for a 60-cm object distance.

Use your diagram to confirm that that the image is behind the mirror, upright, virtual and diminished.   ◊

15. A spherical mirror is to be used to form, on a screen located 5.00 m from the object, an image five times the size of the object. (a) Describe the type of mirror required. (b) Where should the mirror be positioned relative to the object?

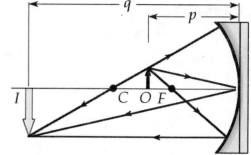

Solution

We are given that $\qquad q = (p + 5.00\text{ m}),\qquad M = -\dfrac{q}{p}\qquad$ and $\qquad |M| = 5.00$

Since the image must be real, q must be positive, and $\qquad\qquad M = -5.00$

or $\qquad\qquad q = 5.00p$

(b) Solving for the distance of the mirror from the object, $\qquad p + 5.00 = 5.00p$

$$p = 1.25\text{ m} \qquad\qquad \Diamond$$

(a) Applying Equation 26.6,

$$\frac{1}{f} = \frac{1}{p} + \frac{1}{q}$$

$$\frac{1}{f} = \frac{1}{1.25\text{ m}} + \frac{1}{6.25\text{ m}}$$

so that the focal length of the mirror is $\qquad\qquad f = 1.04\text{ m}$

Noting that the image is real, inverted and enlarged, we can say that the mirror must be concave, and must have a radius of curvature of $R = 2f = 2.08$ m $\qquad \Diamond$

---

21. A glass sphere ($n = 1.50$) with a radius of 15.0 cm has a tiny air bubble 5.00 cm above its center. The sphere is viewed looking down along the extended radius containing the bubble. What is the apparent depth of the bubble below the surface of the sphere?

Solution

From Equation 26.8,

$$\frac{n_1}{p} + \frac{n_2}{q} = \frac{n_2 - n_1}{R}$$

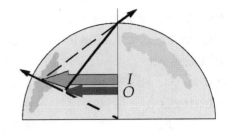

We solve for q to find

$$q = \frac{n_2 R p}{p(n_2 - n_1) - n_1 R}$$

In this case,

$$n_1 = 1.50 \qquad n_2 = 1.00 \qquad p = 10.0 \text{ cm}$$

So the apparent depth is

$$q = \frac{(1.00)(-15.0 \text{ cm})(10.0 \text{ cm})}{(10.0 \text{ cm})(1.00 - 1.50) - (1.50)(-15.0 \text{ cm})} = -8.57 \text{ cm} \ \Diamond$$

To sketch a ray diagram we use the wave under refraction model, as shown. The image is virtual, upright, and enlarged.

---

23. The left face of a biconvex lens has a radius of curvature of magnitude 12.0 cm, and the right face has a radius of curvature of magnitude 18.0 cm. The index of refraction of the glass is 1.44. (a) Calculate the focal length of the lens. (b) Calculate the focal length if the radii of curvature of the two faces are interchanged.

Solution   For a biconvex lens (convex surface on both sides) the center of curvature of each surface is on the opposite side of the lens. The second surface has negative radius:

(a) $\dfrac{1}{f} = (n-1)\left(\dfrac{1}{R_1} - \dfrac{1}{R_2}\right) = (1.44 - 1.00)\left[\dfrac{1}{12.0 \text{ cm}} - \dfrac{1}{(-18.0 \text{ cm})}\right]$

  `18.0 cm          12.0 cm`

$f = 16.4 \text{ cm}$ $\qquad\qquad\qquad\qquad\qquad\qquad\qquad\qquad \Diamond$

(b) $\dfrac{1}{f} = (0.440)\left[\dfrac{1}{18.0 \text{ cm}} - \dfrac{1}{(-12.0 \text{ cm})}\right]$ $\qquad\qquad f = 16.4 \text{ cm} \qquad\qquad \Diamond$

Reversing the lens does not change what it does to the light.

---

27. The nickel's image in Figure P26.27 has twice the diameter of the nickel and is 2.84 cm from the lens. Determine the focal length of the lens.

Solution Looking through the lens, you see the image beyond the lens. Therefore, the image is virtual, with $q = -2.84$ cm.

Now, $$M = \frac{h'}{h} = 2 = -\frac{q}{p},$$

so $$p = -\frac{q}{2} = 1.42 \text{ cm}$$

Thus, $$f = \left(\frac{1}{p} + \frac{1}{q}\right)^{-1} = \left[\frac{1}{1.42 \text{ cm}} + \frac{1}{(-2.84 \text{ cm})}\right]^{-1}$$

and $$f = 2.84 \text{ cm}$$ ◊

33. An object is positioned 20.0 cm to the left of a diverging lens with focal length $f = -32.0$ cm. Determine (a) the location and (b) the magnification of the image. (c) Construct a ray diagram for this arrangement.

Solution

$$\frac{1}{p} + \frac{1}{q} = \frac{1}{f} \quad \text{or} \quad \frac{1}{20.0 \text{ cm}} + \frac{1}{q} = \frac{1}{-32.0 \text{ cm}}$$

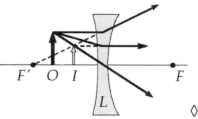

(a) So $$q = -\left(\frac{1}{20.0 \text{ cm}} + \frac{1}{32.0 \text{ cm}}\right)^{-1} = -12.3 \text{ cm}$$ ◊

(b) $$M = -\frac{q}{p} = -\frac{(-12.3 \text{ cm})}{20.0 \text{ cm}} = 0.615$$ ◊

(c) The image is virtual, upright, diminished; the ray diagram is given above. ◊

49. A parallel beam of light enters a glass hemisphere perpendicular to the flat face, as shown in Figure P26.49. The magnitude of the radius is 6.00 cm, and the index of refraction is $n = 1.560$. Determine the point at which the beam is focused. (Assume paraxial rays.)

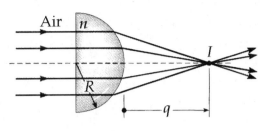

Figure P26.49

Solution   A hemisphere is too thick to be described as a thin lens. The light is undeviated on entry into the flat face. Because of this, we instead consider the light's exit from the second surface, for which $R = -6.00$ cm. The incident rays are parallel, so $p = \infty$.

Then $\dfrac{n_1}{p} + \dfrac{n_2}{q} = \dfrac{n_2 - n_1}{R}$ becomes $0 + \dfrac{1}{q} = \dfrac{1 - 1.56}{-6.00 \text{ cm}}$

and $q = 10.7$ cm   ◊

---

51. The disk of the Sun subtends an angle of 0.533° at the Earth. What are the position and diameter of the solar image formed by a concave spherical mirror with a radius of curvature of 3.00 m?

Solution

For the mirror, $f = R/2 = +1.50$ m. In addition, because the distance to the Sun is so much larger than any other distances, we can take $p = \infty$.

The mirror equation, $\dfrac{1}{p} + \dfrac{1}{q} = \dfrac{1}{f}$, then gives a distance from the mirror of $q = f = 1.50$ m   ◊

Now, in $M = -q/p = h'/h$, the magnification is nearly zero, but we can be more precise: the definition of radian measure means that h/p is the angular diameter of the object. Thus the image diameter is

$$h' = -\dfrac{hq}{p} = (-0.533°)\left(\dfrac{\pi}{180} \text{ rad/deg}\right)(1.50 \text{ m}) = -0.140 \text{ m} = -1.40 \text{ cm}   ◊$$

---

53. An object is placed 12.0 cm to the left of a diverging lens with a focal length of –6.00 cm. A converging lens with a focal length of 12.0 cm is placed a distance d to the right of the diverging lens. Find the distance d that corresponds to a final image at infinity. Draw a ray diagram for this case.

Solution

From Equation 26.12,

$$q_1 = \frac{f_1 p_1}{p_1 - f_1}$$

$$q_1 = \frac{(-6.00 \text{ cm})(12.0 \text{ cm})}{12.0 \text{ cm} - (-6.00 \text{ cm})} = -4.00 \text{ cm}$$

When we require that

$$q_2 \to \infty$$

Equation 26.12 becomes

$$p_2 = f_2 = 12.0 \text{ cm}$$

Since the object for the converging lens must be 12.0 cm to its left, and since this is the image for the diverging lens which is 4.00 cm to **its** left, the two lenses must be separated by 8.00 cm. Mathematically,

$$p_2 = d - (-4.00 \text{ cm})$$

$$d + 4.00 \text{ cm} = f_2 = 12.0 \text{ cm}$$

and

$$d = 8.00 \text{ cm} \qquad \Diamond$$

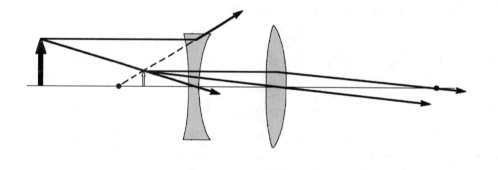

55. In a darkened room, a burning candle is placed 1.50 m from a white wall. A lens is placed between candle and wall at a location that causes a larger, inverted image to form on the wall. When the lens is moved 90.0 cm toward the wall, another image of the candle is formed. Find (a) the two object distances that produce the specified images and (b) the focal length of the lens. (c) Characterize the second image.

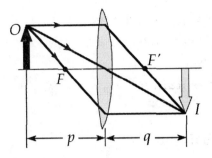

Solution

Originally, $\quad\quad\quad\quad\quad\quad$ $p_1 + q_1 = 1.50$ m

In the final situation, $\quad\quad$ $p_2 = p_1 + 0.900$ m,

And $\quad\quad\quad\quad\quad\quad\quad$ $q_2 = q_1 - 0.900$ m $= (1.50$ m $- p_1) - 0.900$ m $= 0.600$ m $- p_1$

Our lens equation is $\quad\quad$ $\dfrac{1}{p_1} + \dfrac{1}{q_1} = \dfrac{1}{f} = \dfrac{1}{p_2} + \dfrac{1}{q_2}$

Substituting, we have $\quad\quad$ $\dfrac{1}{p_1} + \dfrac{1}{1.50 \text{ m} - p_1} = \dfrac{1}{p_1 + 0.900} + \dfrac{1}{0.600 - p_1}$

Adding the fractions, $\quad\quad$ $\dfrac{1.50 \text{ m} - p_1 + p_1}{p_1(1.50 \text{ m} - p_1)} = \dfrac{0.600 - p_1 + p_1 + 0.900}{(p_1 + 0.900)(0.600 - p_1)}$

Simplified, this becomes $\quad$ $p_1(1.50 \text{ m} - p_1) = (p_1 + 0.900)(0.600 - p_1)$

(a)   Thus, $\quad\quad\quad\quad\quad\quad$ $p_1 = \dfrac{0.540}{1.80}$ m $= 0.300$ m $\quad\quad\quad\quad\quad\quad\quad$ ◊

$\quad\quad\quad\quad\quad\quad\quad\quad\quad$ $p_2 = p_1 + 0.900$ m $= 1.20$ m $\quad\quad\quad\quad\quad\quad\quad$ ◊

(b) $\dfrac{1}{f} = \dfrac{1}{0.300 \text{ m}} + \dfrac{1}{1.50 \text{ m} - 0.300 \text{ m}}$ $\quad\quad$ and $\quad\quad$ $f = 0.240$ m $\quad\quad$ ◊

(c)   The second image is real, inverted, and diminished, with $\quad\quad$ $M = \dfrac{-q_2}{p_2} = -0.250$ $\quad$ ◊

Chapter 27

Wave Optics

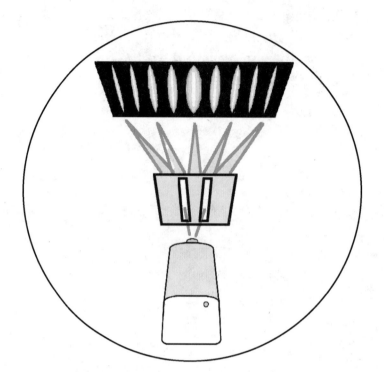

INTRODUCTION

In the previous chapter on geometric optics, we used light rays to explain the formation of images when light passes through a lens or reflects from a mirror. This chapter is concerned with wave optics, which deals with the interference and diffraction. These phenomena cannot be adequately explained with the ray optics of Chapter 26. However, by treating light as waves rather than as rays, we can give a satisfactory description of such phenomena. In using the wave in interference model we will construct geometric models in two or three dimensions and apply the superposition principle to the addition of electromagnetic fields.

NOTES FROM SELECTED CHAPTER SECTIONS

27.1   Conditions for Interference

In order to observe **sustained interference** in light waves, the following conditions must be met:
- The sources must be coherent; they must maintain a **constant phase** with respect to each other.
- The sources must be **monochromatic** – of a **single wavelength.**

27.2   Young's Double-Slit Experiment

A schematic diagram illustrating the geometry used in Young's double-slit experiment is shown in the figure below. The two slits S_1 and S_2 serve as coherent monochromatic sources. When L (the distance from source plane to viewing screen) is much greater than d (the distance between sources), the **path difference** $\delta = r_2 - r_1 = d\sin\theta.$

27.4   Change of Phase due to Reflection

An electromagnetic wave undergoes a **phase change of 180°** upon reflection from a medium that has **a larger index of refraction** than the one in which it is traveling. There is also a 180° phase change upon reflection from a **conducting surface.**

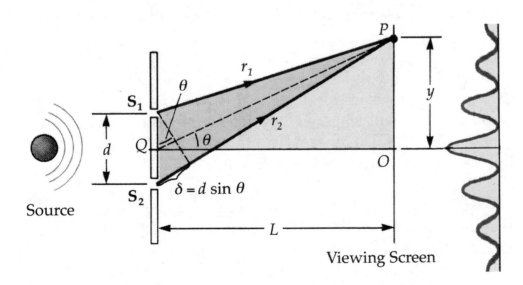

Figure 27.1

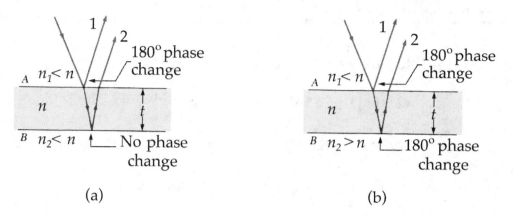

Figure 27.2 (a) Interference of light resulting from reflections at two surfaces of a thin film of thickness t, index of refraction n. (b) A thin film between two different media.

27.5 Interference in Thin Films

Interference effects in thin films depend on the difference in length of path traveled by the interfering waves as well as any phase changes which may occur due to reflection. It can be shown by the Lloyd's mirror experiment that when light is reflected from a surface of index of refraction greater than the index of the medium in which it is initially traveling, the reflected ray undergoes a 180° (or π radians) phase change. In analyzing interference effects, this can be considered equivalent to the gain or loss of a half wavelength in path difference. Therefore, there are two different cases to consider: (a) a film surrounded by a common medium and (b) a thin film located between two different media. These cases are illustrated in Figure 27.2 above.

In case (a), where a phase change occurs at the top surface, the reflected rays will be in phase (constructive interference) if the thickness of the film, t, is an odd number of quarter wavelengths in the film – so that the path lengths will differ by an odd number of half wavelengths as measured in the film. NOTE: the **wavelength in the film**, $\lambda_n = \lambda / n$ where λ = wavelength of the light in free space and n = index of refraction of the film

In case (b), the phase changes due to reflection at **both** the top and bottom surfaces are offsetting and, therefore, constructive interference for the reflected rays will occur when the film thickness is an integer number of half wavelengths (as measured in the film) – and, therefore, the path difference will be a whole number of wavelengths (measured in the film).

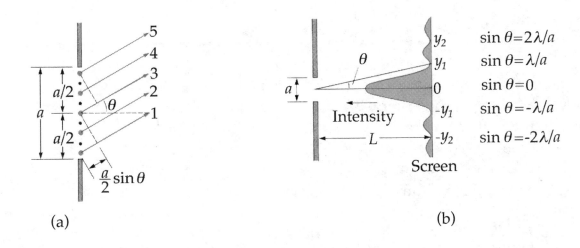

Figure 27.3 (a) Diffraction of light by a narrow slit of width a, and (b) the resulting intensity distribution.

27.6 Diffraction Patterns

Fraunhofer diffraction occurs when light rays reaching a point (on a screen) from a diffracting source (edge, slit, etc.) are approximately parallel. According to **Huygens's principle**, each portion of a slit acts as a source of waves; therefore, light from one portion of a slit can interfere with light from another portion.

In Figure 27.3 (a) above, parallel light rays are shown incident on a slit of width a. In order to analyze the resulting Fraunhofer diffraction pattern, it is convenient to subdivide the width of the slit into many strips of equal width. Each strip can be considered as a source of waves. The positions of the minima in a single-slit diffraction pattern are determined by the conditions stated in Equation 27.14. The **secondary maxima** (of progressively diminished intensity) **lie approximately halfway between the minima.**

27.7 Resolution of Single-Slit and Circular Apertures

When the central maximum of one image falls on the first minimum of another image, the images are said to be just resolved. This limiting condition of resolution is known as **Rayleigh's criterion.**

EQUATIONS AND CONCEPTS

In the arrangement used for the Young's double-slit experiment (Figure 27.1), two slits separated by a distance, d, serve as monochromatic coherent sources. The light intensity at any point on the screen is the resultant of light reaching the screen from both slits. When using $y = L\tan\theta$, y is measured from the center of the interference pattern and L is the distance from the double slit to the screen. Light from the two slits reaching any point on the screen (except the center) travels unequal path lengths. This difference in length of path, δ, is called the path difference.

$$\delta = r_2 - r_1 = d\sin\theta \qquad (27.1)$$

$$y = L\tan\theta$$

Bright fringes (constructive interference) will appear at points on the screen for which the path difference is equal to an integral multiple of the wavelength. The positions of bright fringes can also be located by calculating their distance from the center of the screen (y). In each case, the number m is called the order number of the fringe. The central bright fringe ($\theta = 0$, $m = 0$) is called the zeroth-order maximum.

$$\delta = d\sin\theta_{bright} = \lambda m \qquad (27.2)$$
$$\text{for} \quad m = 0, \ \pm 1, \ \pm 2, \ \ldots$$

$$y_{bright} \cong m\frac{\lambda L}{d} \qquad (27.5)$$
$$\text{(for small } \theta)$$

Dark fringes (destructive interference) will appear at points on the screen, which correspond to path differences of an odd multiple of half wavelengths. For these points of destructive interference, waves which leave the two slits in phase arrive at the screen 180° out of phase.

$$\delta = d\sin\theta_{dark} = \left(m + \frac{1}{2}\right)\lambda \qquad (27.3)$$
$$\text{for} \quad m = 0, \ \pm 1, \ \pm 2, \ \ldots$$

$$y_{dark} \cong \left(m + \frac{1}{2}\right)\frac{\lambda L}{d} \qquad (27.6)$$
$$\text{(for small } \theta)$$

Two waves which leave the slits initially in phase arrive at the screen **out of phase** by an amount, which depends on the path difference.

$$\phi = \frac{2\pi}{\lambda} d \sin\theta \qquad (27.7)$$

$$\phi = \frac{2\pi}{\lambda} \delta$$

The average light intensity (I_{av}) at any point P on the screen is proportional to the square of the amplitude of the resultant wave. **The average intensity can be written:**

- as a function of phase difference ϕ;

$$I_{av} = I_{max} \cos^2\left(\frac{\phi}{2}\right)$$

- as a function of the angle (θ) subtended by the screen point at the source midpoint; or

$$I_{av} = I_{max} \cos^2\left(\frac{\pi d \sin\theta}{\lambda}\right) \qquad (27.8)$$

- as a function of the distance (y) from the center of the screen.

$$I_{av} \cong I_{max} \cos^2\left(\frac{\pi d}{\lambda L} y\right) \qquad (27.9)$$

(for small θ)

The intensity pattern observed on the screen will vary as the number of equally spaced sources is increased; however, the **positions of the principal maxima remain the same.**

In thin-film interference (Figure 27.2), the wavelength of light in the film, λ_n, is the ratio of the wavelength in free space to the index of refraction of the film.

$$\lambda_n = \frac{\lambda}{n} \qquad (27.10)$$

The conditions for the interference of light reflected perpendicularly from a thin film can be stated in terms of the thickness (t) and the index of refraction of the film. The conditions expressed by Equations 27.12 and 27.13 are valid when:

(1) The film is surrounded by a common medium (see Figure 27.2a) with an index either less than or greater than that of the film; or

(2) different media are above and below the film and both media have indices which are greater than or less than the index of the film.

Figure 27.2b illustrates the case when the index of refraction of the film is greater than that of the medium above the film and less than the index of the medium below the film. In this case a phase change of 180° occurs upon reflection at both the top and bottom surfaces of the film. Under these conditions, constructive interference will be observed when the film thickness is an odd number of half-wavelengths as measured in the film.

The general **condition for destructive interference** in the diffraction pattern of a single slit in terms of θ is a function of the slit width, a, and wavelength of incident light.

$$2nt = \left(m + \frac{1}{2}\right)\lambda \quad (m = 0, 1, 2, \ldots) \quad (27.12)$$
constructive interference

$$2nt = m\lambda \quad (m = 0, 1, 2, \ldots) \quad (27.13)$$
destructive interference

$$\sin\theta_{\text{dark}} = m\frac{\lambda}{a} \quad (27.14)$$
$$(m = \pm1, \pm2, \pm3, \ldots)$$
(Destructive interference)

Rayleigh's criterion states the condition for the resolution of two images due to nearby sources. For a slit, the angular separation between the sources must be greater than the ratio of the wavelength to slit width if the two sources are to be resolved. In the case of a **circular aperture**, the minimum angular separation depends on D, the diameter of the aperture (or lens).

$$\theta_{min} = \frac{\lambda}{a} \qquad (27.15)$$
$$\text{(slit)}$$

$$\theta_{min} = 1.22\frac{\lambda}{D} \qquad (27.16)$$
$$\text{(circular aperture)}$$

If light containing several wavelengths is incident on a grating, a series of spectra (each corresponding to an order number, m) will be observed. Each spectrum will contain a line characteristic of each wavelength. Equation 27.17 gives the angle of deviation for constructive interference (bright lines in the spectrum).

$$d\sin\theta_{bright} = m\lambda \qquad (27.17)$$
$$(m = 0, 1, 2, 3, \ldots)$$

Bragg's law gives the conditions for **constructive interference** of x-rays reflected from the parallel planes of a crystalline solid separated by a distance d. **The angle θ is the angle between the incident beam and the surface.**

$$2d\sin\theta = m\lambda \qquad (27.18)$$
$$(m = 1, 2, 3, \ldots)$$

SUGGESTIONS, SKILLS, AND STRATEGIES

The following features should be kept in mind while working thin-film interference problems:

- Identify the thin film from which interference effects are being observed.
- The type of interference that occurs in a specific problem is determined by the phase relationship between that portion of the wave reflected at the upper surface of the film and that portion reflected at the lower surface of the film.
- Phase differences between the two portions of the wave occur because of differences in the distances traveled by the two portions and by phase changes occurring upon reflection.

The wave reflected from the lower surface of the film has to travel a distance equal to twice the thickness of the film before it returns to the upper surface of the film. When **distances alone are considered**, if the extra distance is equal to an integral multiple of λ_n constructive interference will occur; if the extra distance equals $\frac{1}{2}\lambda_n$, $\frac{3}{2}\lambda_n$, and so forth, destructive interference will occur.

Reflections may change such results. When a traveling wave reflects off a surface having a higher index of refraction than the one it is in, a 180° phase shift occurs. This has the same effect as if the wave lost $\frac{1}{2}\lambda_n$. These shifts must be considered in addition to shifts that occur because of the extra distance that one wave travels.

- When both distance and phase changes upon reflection are taken into account, the interference will be constructive if the waves are out of phase by an integral multiple of λ_n. Destructive interference will occur when the total phase difference corresponds to $\frac{1}{2}\lambda_n$, $\frac{3}{2}\lambda_n$, and so forth. For example:

In a soap bubble (Fig. 27.2a), $n_{air} < n_{soap} > n_{air}$, the condition for constructive interference is

$$\Delta d = \left(m + \frac{1}{2}\right)\lambda_n.$$

In an oil slick (Fig. 27.2b), $n_{air} < n_{oil} < n_{asphalt}$, the condition for constructive interference is

$$\Delta d = m\,\lambda_n.$$

REVIEW CHECKLIST

▷ Describe Young's double-slit experiment to demonstrate the wave nature of light. Account for the phase difference between light waves from the two sources as they arrive at a given point on the screen. State the conditions for constructive and destructive interference in terms of each of the following: path difference, phase difference, distance from the center of the screen, and angle subtended by the observation point at the source mid-point.

▷ Outline the manner in which the superposition principle leads to the correct expression for the intensity distribution on a distant screen due to two coherent sources of equal intensity.

▷ Account for the conditions of constructive and destructive interference in thin films considering both path difference and any expected phase changes due to reflection.

▷ Determine the positions of the minima in a single-slit diffraction pattern. Determine the positions of the principal maxima in the interference pattern of a diffraction grating.

▷ Determine whether or not two sources under a given set of conditions are resolvable as defined by Rayleigh's criterion.

ANSWERS TO SELECTED CONCEPTUAL QUESTIONS

1.   What is the necessary condition on the path length difference between two waves that interfere (a) constructively and (b) destructively?

Answer

(a) Two waves interfere constructively if their path difference is either zero or some integral multiple of the wavelength; that is, if the path difference equals $m\lambda$. (b) Two waves interfere destructively if their path difference is an odd multiple of one-half of a wavelength; that is, if the path difference equals $\left(m+\frac{1}{2}\right)\lambda$.

4. In Young's double-slit experiment, why do we use monochromatic light? If white light were used, how would the pattern change?

Answer

Every color produces its own pattern, with a spacing between the maxima that is characteristic of the wavelength. With several colors, the patterns are superimposed, and it can become difficult to pick out a single maximum. Using monochromatic light can eliminate this problem.

With white light, the central maximum is white. The first side maximum is a full spectrum, with violet on the inside and red on the outside. The second side maximum is a full spectrum also, but red in the second maximum overlaps the violet in the third maximum. At larger angles, the light soon starts mixing to white again, though it is often so faint that you would call it gray.

12. Would it be possible to place a nonreflective coating on an airplane to cancel radar waves of wavelength 3 cm?

Answer

Yes. In order to do this, first measure the radar-reflectivity of the metal of your airplane. Then, choose a light, durable material that has approximately half the radar-reflectivity of the metal in your plane. Measure its index of refraction, and onto the metal plaster a coating equal in thickness to one-quarter of 3 cm, divided by that index.

This will effectively minimize the radar reflection at 3 cm; however, it will maximize the reflection at 1.5 cm. Thus, while it is possible to do this, it is not effective in a stealth design.

15. Although we can hear around corners, we cannot see around corners. How can you explain this in view of the fact that sound and light are both waves?

Answer

Audible sound has wavelengths on the order of meters or centimeters, while visible light has wavelengths on the order of half a micron. In this world of breadbox-size objects, λ is comparable to the object size a for sound, and sound diffracts around walls and through doorways. But λ/a is much smaller for visible light passing ordinary-size objects or apertures, so light diffracts only through very small angles.

Another way of answering this question would be as follows. We can see by a small angle around a small obstacle or around the edge of a small opening. The side fringes in Figure 27.12 and the Arago spot in the center of Figure 27.13 (in the textbook) show this diffraction. Conversely, we cannot always hear around corners. Out-of-doors, away from reflecting surfaces, have someone a few meters distant face away from you and whisper. The high-frequency, short-wavelength, information-carrying components of the sound do not diffract around his head enough for you to understand his words.

□ □ □ □

20. Describe the change in width of the central maximum of the single-slit diffraction pattern as the width of the slit is made smaller.

Answer

Equation 27.14 describes the angles at which you get destructive interference; from it, we can obtain an estimate of the width of the central maximum. For small angles, the equation can be rewritten as

$$\theta_m = \sin^{-1}\left(\frac{m\lambda}{a}\right) \cong \frac{m\lambda}{a}$$

Thus, as the width of the slit a decreases, the angle of the first destructive interference θ_1 grows, and the width of the central maximum grows as well.

SOLUTIONS TO SELECTED END–OF–CHAPTER PROBLEMS

3. Two radio antennas separated by 300 m, as in Figure P27.3 simultaneously broadcast identical signals at the same wavelength. A radio in a car traveling due north receives the signals. (a) If the car is at the position of the second maximum, what is the wavelength of the signals? (b) How much farther must the car travel to encounter the next minimum in reception? (**Note:** Do not use the small-angle approximation in this problem.)

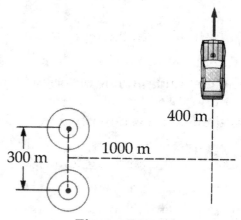

Figure P27.3

Solution Note that with the conditions given, the small angle approximation **does not work well.** That is, $\sin\theta$, $\tan\theta$, and θ are significantly different. The approach to be used is outlined below. We use the waves in interference model.

(a) At the $m = 2$ maximum,

$$\tan\theta = \frac{400 \text{ m}}{1000 \text{ m}} = 0.400$$

and

$$\theta = 21.8°$$

so

$$\lambda = \frac{d\sin\theta}{m} = \frac{(300 \text{ m})\sin 21.8°}{2} = 55.7 \text{ m} \qquad \Diamond$$

(b) The next minimum encountered is the $m = 2$ minimum.

At that point $d\sin\theta = \left(m + \frac{1}{2}\right)\lambda$ $\qquad\qquad d\sin\theta = \frac{5}{2}\lambda$

or $\sin\theta = \frac{5\lambda}{2d} = \frac{5(55.7 \text{ m})}{2(300 \text{ m})} = 0.464$ $\qquad\qquad \theta = 27.7°$

so $y = (1000 \text{ m})\tan 27.7° = 524 \text{ m}$

Therefore, the car must travel an additional 124 m. $\qquad \Diamond$

5. Young's double-slit experiment is performed with 589-nm light and distance of 2.00 m between the slits and the screen. The tenth interference minimum is observed 7.26 mm from the central maximum. Determine the spacing of the slits.

Solution   In the equation $d \sin\theta = \left(m + \frac{1}{2}\right)\lambda$

the first minimum is described by $m = 0$

and the tenth is described by $m = 9$

Thus, for the tenth minimum, $\sin\theta = \frac{\lambda}{d}\left(9 + \frac{1}{2}\right)$

For small θ, $\sin\theta \cong \tan\theta$, and $\tan\theta = \frac{y}{L}$

Solving for d,
$$d = \frac{9.5\lambda}{\sin\theta} = \frac{9.5\lambda L}{y} = \frac{9.5\left(5890 \times 10^{-10}\text{ m}\right)(2.00\text{ m})}{7.26 \times 10^{-3}\text{ m}}$$

$$d = 1.54 \times 10^{-3}\text{ m} = 1.54\text{ mm} \qquad \Diamond$$

---

9. In Figure 27.1 of this solution manual, let $L = 120$ cm and $d = 0.250$ cm. The slits are illuminated with coherent 600-nm light. Calculate the distance y above the central maximum for which the intensity on the screen is 75.0% of the maximum.

Solution   For small θ, $I_{av} = I_{max}\cos^2\left(\frac{\pi d \sin\theta}{\lambda}\right)$

From the drawing, $\sin\theta \cong \frac{y}{L}$: $y = \frac{\lambda L}{\pi d}\cos^{-1}\sqrt{\frac{I_{av}}{I_{max}}}$

In addition, since $I_{av} = 0.750\ I_{max}$

we can substitute a value for each variable: $y = \frac{\left(6.00 \times 10^{-7}\text{ m}\right)(1.20\text{ m})}{\pi\left(2.50 \times 10^{-3}\text{ m}\right)}\cos^{-1}\sqrt{0.750}$

$$y = 0.0480\text{ mm} \qquad \Diamond$$

13. An oil film ($n = 1.45$) floating on water is illuminated by white light at normal incidence. The film is 280 nm thick. Find (a) the dominant observed color in the reflected light and (b) the dominant color in the transmitted light. Explain your reasoning.

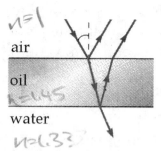

Solution

The light reflected from the top of the oil film undergoes phase reversal. Since $1.45 > 1.33$, the light reflected from the bottom undergoes no reversal. For constructive interference of reflected light, we then have

$$2nt = \left(m + \tfrac{1}{2}\right)\lambda \qquad \text{or} \qquad \lambda_m = \frac{2nt}{m + \dfrac{1}{2}} = \frac{2(1.45)(280 \text{ nm})}{m + \dfrac{1}{2}}$$

(a) Substituting for m, we have

$m = 0$: $\lambda_0 = 1624$ nm (infrared)
$m = 1$: $\lambda_1 = 541$ nm (green)
$m = 2$: $\lambda_2 = 325$ nm (ultraviolet)

Both infrared and ultraviolet light are invisible to the human eye, so the dominant color is green.

(b) Any light that is not reflected is transmitted. To find the color transmitted most strongly, we find the wavelengths reflected least strongly. According to the condition for destructive interference $2nt = m\lambda$.

Therefore,
$$\lambda = \frac{2nt}{m} = \frac{2(1.45)(280 \text{ nm})}{m}$$

For $m = 1$, $\lambda = 812$ nm   (near infrared)
             $m = 2$, $\lambda = 406$ nm   (violet)
             $m = 3$,   (ultraviolet)

Thus violet is the only visible color not attenuated by reflection, and the dominant color in the transmitted light.

15. An air wedge is formed between two glass plates separated at one edge by a very fine wire as in Figure P27.15. When the wedge is illuminated from above by 600-nm light, 30 dark fringes are observed. Calculate the radius of the wire.

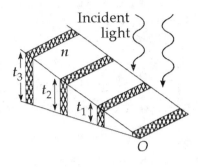

Figure P27.15

Solution   Light reflecting from the bottom surface of the top plate undergoes no phase shift, while light reflecting from the top surface of the bottom plate is shifted by π, and must also travel an extra distance $2t$, where t is the thickness of the air wedge.

For destructive interference, $2t = m\lambda$ (where $m = 0, 1, 2, 3, \ldots$). The first dark fringe appears where $m = 0$ at the line of contact between the plates. The thirtieth dark fringe gives for the diameter of the wire

$$2t = 29\lambda, \qquad \text{so} \qquad t = 14.5\lambda$$

$$r = \frac{1}{2}t = \frac{1}{2}(14.5)\left(600 \times 10^{-9} \text{ m}\right) = 4.35 \ \mu\text{m} \qquad \Diamond$$

19. A screen is placed 50.0 cm from a single slit, which is illuminated with 690-nm light. If the distance between the first and third minima in the diffraction pattern is 3.00 mm, what is the width of the slit?

Solution   In the equation for single-slit diffraction minima at small angles

(Eq. 27.14)
$$\frac{y}{L} \cong \sin\theta_{\text{dark}} = \frac{m\lambda}{a}$$

take differences between the first and third minima, to see that $\dfrac{\Delta y}{L} = \dfrac{\Delta m \lambda}{a}$

with $\Delta y = 3.00 \times 10^{-3}$ m and $\Delta m = 3 - 1 = 2$

The width of the slit is then
$$a = \frac{\lambda L \Delta m}{\Delta y} = \frac{\left(690 \times 10^{-9} \text{ m}\right)(0.500 \text{ m})(2)}{3.00 \times 10^{-3} \text{ m}}$$

$$a = 2.30 \times 10^{-4} \text{ m} \qquad \Diamond$$

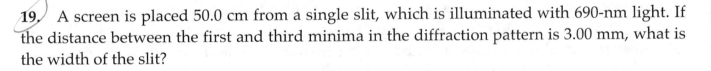

23.   A helium-neon laser emits light that has a wavelength of 632.8 nm. The circular aperture through which the beam emerges has a diameter of 0.500 cm. Estimate the diameter of the beam 10.0 km from the laser.

Solution   Following Equation 27.16 for diffraction from a circular opening, the beam spreads into a cone of half-angle

$$\theta_{min} = 1.22\frac{\lambda}{D} = 1.22\left(\frac{632.8\times10^{-9}\ \text{m}}{0.00500\ \text{m}}\right) = 1.54\times10^{-4}\ \text{rad}$$

The radius of the beam ten kilometers away is, from the definition of radian measure,

$$r_{beam} = \theta_{min}\left(1.00\times10^4\ \text{m}\right) = 1.54\ \text{m}$$

and its diameter is $d_{beam} = 2r_{beam} = 3.09\ \text{m}$ ◊

---

25. The Impressionist painter Georges Seurat created paintings with an enormous number of dots of pure pigment, each of which was approximately 2.00 mm in diameter. The idea was to have colors such as red and green next to each other to form a scintillating canvas (Fig. P27.25). Outside what distance would one be unable to discern individuals dots on the canvas? (Assume $\lambda = 500$ nm and that the pupil diameter is 4.00 mm.)

Solution   We will assume that the dots are just touching and do not overlap, so that the distance between their centers is 2.00 mm. By Rayleigh's criterion, two dots separated center to center by 2.00 mm would be seen to overlap when

$$\theta_{min} = \frac{d}{L} = 1.22\frac{\lambda}{D}$$

with $d = 2.00$ mm, $\lambda = 500$ nm, and $D = 4.00$ mm

Thus, $L = \dfrac{Dd}{1.22\lambda} = \dfrac{\left(4.00\times10^{-3}\ \text{m}\right)\left(2.00\times10^{-3}\ \text{m}\right)}{1.22\left(500\times10^{-9}\ \text{m}\right)} = 13.1\ \text{m}$ ◊

31. The hydrogen spectrum has a red line at 656 nm and a blue line at 434 nm. What are the angular separations between two spectral lines obtained with a diffraction grating that has 4500 lines/cm?

Solution

The grating spacing is
$$d = \frac{1.00 \times 10^{-2} \text{ m}}{4500} = 2.22 \times 10^{-6} \text{ m}$$

In the first-order spectrum, the angles of diffraction are given by $\sin \theta = \lambda / d$:

$$\sin \theta_1 = \frac{656 \times 10^{-9} \text{ m}}{2.22 \times 10^{-6} \text{ m}} = 0.295$$

$$\sin \theta_2 = \frac{434 \times 10^{-9} \text{ m}}{2.22 \times 10^{-6} \text{ m}} = 0.195$$

so that $\theta_1 = 17.17°$ and $\theta_2 = 11.26°$

The angular separation is therefore $\Delta\theta = 17.17° - 11.26° = 5.91°$ ◊

In the second-order spectrum, $\Delta\theta = \sin^{-1}\left(\frac{2\lambda_1}{d}\right) - \sin^{-1}\left(\frac{2\lambda_2}{d}\right) = 13.2°$ ◊

Again, in the third order, $\Delta\theta = \sin^{-1}\left(\frac{3\lambda_1}{d}\right) - \sin^{-1}\left(\frac{3\lambda_2}{d}\right) = 26.5°$ ◊

Since the red line does not appear in the fourth-order spectrum, the answer is complete.

---

35. A grating with 250 lines/mm is used with an incandescent light source. Assume the visible spectrum to range in wavelength from 400 to 700 nm. In how many orders can one see (a) the entire visible spectrum and (b) the short-wavelength region?

Solution   The grating spacing is $d = \dfrac{1.00 \text{ mm}}{250} = 4.00 \times 10^{-6}$ m

(a) In each order of interference m, red light diffracts at a larger angle than the other colors with shorter wavelengths. We find the largest integer m satisfying

$$d \sin \theta = m \lambda \qquad \text{with} \qquad \lambda = 700 \text{ nm}$$

With $\sin \theta$ having its largest value $\left(4.00 \times 10^{-6} \text{ m}\right) \sin 90° = m\left(700 \times 10^{-9} \text{ m}\right)$

so that $m = 5.71$

Thus the red light cannot be seen in the 6th order, and the full visible spectrum appears in only five orders.   ◊

(b) Now consider light at the boundary between violet and ultraviolet.

$d \sin \theta = m \lambda$ becomes $\left(4.00 \times 10^{-6} \text{ m}\right) \sin 90° = m\left(400 \times 10^{-9} \text{ m}\right)$

and $m = 10$ ◊

---

37. If the interplanar spacing of NaCl is 0.281 nm, what is the predicted angle at which 0.140-nm x-rays are diffracted in a first-order maximum?

Solution   The atomic planes in this crystal are shown in Figure 27.24 of the text. The diffraction they produce is described by the Bragg condition, that

$2d \sin \theta = m \lambda$

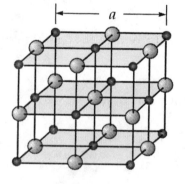

Figure 27.24

$\sin \theta = \dfrac{m \lambda}{2d} = \dfrac{1\left(0.140 \times 10^{-9} \text{ m}\right)}{2\left(0.281 \times 10^{-9} \text{ m}\right)} = 0.249$

$\theta = 14.4°$ ◊

---

47. The condition for constructive interference by reflection from a thin film in air as developed in Section 27.5 assumes nearly normal incidence. Show that if the light is incident on the film at a nonzero angle ϕ_1 (relative to the normal), then the condition for constructive interference is $2nt\cos\theta_2 = \left(m + \frac{1}{2}\right)\lambda$ where θ_2 is the angle of refraction.

Solution

It may be helpful to draw a ray diagram of the interference. Because refraction out of the film is the opposite of refraction into the film, the refracted ray leaves the film parallel to the reflected ray. Then, since the angles marked α are equal,

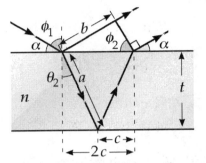

$$\phi_2 + \alpha + 90° = 180°$$

From the angle of incidence, $\phi_1 + \alpha = 90°$

so $\phi_2 = \phi_1$

Before they head off together along parallel paths, one beam travels a distance b and undergoes phase reversal on reflection; the other beam travels a distance $2a$ inside the film, where its wavelength is λ/n.

The number of cycles it completes along this path is $\dfrac{2a}{\lambda/n} = \dfrac{2na}{\lambda}$

The optical path length in the film is $2na$

Then the shift between the two outgoing rays is $\delta = 2na - b - \dfrac{\lambda}{2}$

where a and b are as shown in the ray diagram, n is the index of refraction, and the factor of $\lambda/2$ is due to phase reversal at the top surface. For constructive interference, $\delta = m\lambda$ where m has integer values.

This condition becomes $2na - b = \left(m + \dfrac{1}{2}\right)\lambda$ (1)

From the figure's geometry, $\quad a = \dfrac{t}{\cos\theta_2}$

and $\quad c = a\sin\theta_2 = \dfrac{t\sin\theta_2}{\cos\theta_2}$

Therefore, $\quad b = 2c\sin\phi_1 = \dfrac{2t\sin\theta_2}{\cos\theta_2}\sin\phi_1$

Also, from Snell's law, $\quad \sin\phi_1 = n\sin\theta_2$

so $\quad b = \dfrac{2nt\sin^2\theta_2}{\cos\theta_2}$

With these results, the condition for constructive interference given in Equation (1) becomes:

$$2n\left(\dfrac{t}{\cos\theta_2}\right) - \dfrac{2nt\sin^2\theta_2}{\cos\theta_2} = \dfrac{2nt}{\cos\theta_2}\left(1-\sin^2\theta_2\right) = \left(m+\tfrac{1}{2}\right)\lambda$$

or $\quad 2nt\cos\theta_2 = \left(m+\tfrac{1}{2}\right)\lambda \qquad \lozenge$

Chapter 28

Quantum Physics

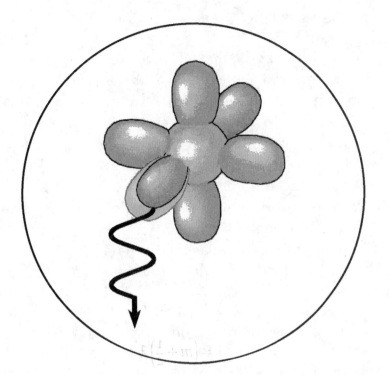

INTRODUCTION

In earlier chapters we introduced two simplification models: the particle model and the wave model. In this chapter we introduce two additional simplification models: the quantum particle model and the quantum particle under boundary conditions model.

Using these models, we introduce the basic principles of quantum mechanics to describe various phenomena including wave-particle duality, blackbody radiation, photoelectric effect, uncertainty principle, behavior of a particle in a box, and Compton scattering.

NOTES FROM SELECTED CHAPTER SECTIONS

28.1 Blackbody Radiation and Planck's Theory

A **black body** is an ideal body that absorbs all radiation incident on it. Any body at some temperature T emits thermal radiation, which is characterized by the properties of the body and its temperature. The spectral distribution of blackbody radiation at various temperatures is shown in the figure. As the temperature increases, the total power of the emitted radiation (area under the curve) increases, while the peak of the distribution shifts to shorter wavelengths.

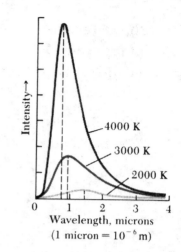

Classical theories failed to explain blackbody radiation. A new theory, proposed by Max Planck, is consistent with this distribution at all wavelengths. Planck made two basic assumptions in the development of this result:

- The oscillators emitting the radiation can only have **discrete energies** given by $E_n = nhf$, where n is a quantum number ($n = 1, 2, 3, \ldots$), f is the oscillator frequency, and h is Planck's constant.
- These oscillators can emit or absorb energy in discrete units called quanta (or photons), where the energy of a light quantum obeys the relation $E = hf$.

Subsequent developments showed that the quantum concept was necessary in order to explain several phenomena at the atomic level, including the photoelectric effect, the Compton effect, and atomic spectra.

28.2 The Photoelectric Effect

When light is incident on certain metallic surfaces, electrons can be emitted from the surfaces. This is called the photoelectric effect, discovered by Hertz. Several features of the photoelectric effect can not be explained with classical physics or with the wave theory of light. In 1905, Einstein provided a successful explanation of the photoelectric effect by extending Planck's quantum concept to include electromagnetic fields.

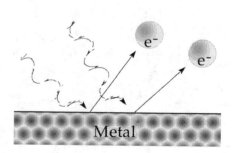

Observed features of the photoelectric effect can be explained and understood on the basis of the photon theory of light. These features include:

- **Photoelectrons have a maximum kinetic energy that is independent of light intensity** (number of photons incident per unit area of the photo surface per second). This is confirmed experimentally by showing that the photocurrent becomes zero at the same stopping potential for different light intensities.

- **Photoelectrons are emitted almost instantaneously** by photons with a frequency above the cutoff frequency (a threshold or minimum value characteristic of a particular material). This instantaneous emission occurs even at very low light intensity.

- **No photoelectrons are emitted when the frequency of incident light is below the cutoff frequency** characteristic of the material being illuminated. This is true regardless of the degree of light intensity.

- **The maximum kinetic energy of photoelectrons increases with increasing light frequency.** This is seen by examining the photoelectric equation (Eq. 28.5), which states that $K_{max} = hf - \phi$. The work function ϕ represents the minimum energy with which an electron is bound in the surface of a metal. The cutoff frequency is $f_c = \phi / h$.

28.3 The Compton Effect

The Compton effect involves the scattering of x-rays by electrons. The scattered x-ray undergoes a **change** in wavelength $\Delta\lambda$, called the Compton shift, which cannot be explained using classical concepts. By treating the x-ray as a photon (the quantum concept), the scattering process between the photon and electron predicts a shift in photon (x-ray) wavelength given by Equation 28.7, where θ is the angle between the incident and scattered x-ray and m_e is the mass of the electron. The formula is in excellent agreement with experimental results.

The figure to the right represents the data for Compton scattering of x-rays from graphite at $\theta = 90.0°$. In this case, the Compton shift is $\Delta\lambda = 0.0024$ nm and λ_0 is the wavelength of the incident x-ray beam.

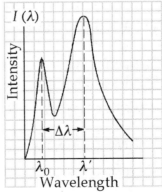

28.4   Photons and Electromagnetic Waves

The results of some experiments are better described based on the particle nature of light; other experimental outcomes are better described in terms of the wave properties of light. Both aspects of the dual nature of light can be described by the **quantum particle model**.

28.5   The Wave Properties of Particles

De Broglie postulated that a particle in motion has wave properties and a corresponding wavelength inversely proportional to the particle's momentum.

28.8   The Uncertainty Principle

It is physically impossible to simultaneously measure the exact position and exact momentum of a particle. The uncertainties in position and momentum are due to the quantum structure of matter.

28.9   An Interpretation of Quantum Mechanics

The wave function is a complex valued quantity, the square of the absolute value of which gives the probability per volume of finding a particle at a given point in the volume; the wave function contains all the information that can be known about the particle.

The **Schrödinger equation** describes the manner in which matter waves change in time and space.

The average experimental value of a quantity such as position or energy is called the **expectation value** of the quantity.

28.10   A Particle in a Box

A particle confined to a line segment and represented by a well-defined de Broglie wave function is represented by a sinusoidal wave. The allowed states of the system are called **stationary states** since they represent **standing waves**.

The minimum energy that the particle can have is called the **zero-point energy**.

28.12 The Schrödinger Equation

The basic problem in wave mechanics is to determine a solution to the Schrödinger equation. The solution will provide the allowed wave functions and energy levels of the system.

28.13 Tunneling Through a Potential Energy Barrier

When a particle is incident onto a barrier, the height of which is greater than the energy of the particle, there is a finite probability that the particle will penetrate the barrier. In this process, called **tunneling**, part of the incident wave is transmitted and part is reflected.

EQUATIONS AND CONCEPTS

The Wien displacement law properly describes the peak in the energy spectrum emitted by a black body radiator. A **black body** is an ideal body that absorbs all radiation incident on it. Any body at some temperature T emits thermal radiation, which is characterized by the properties of the body and its temperature. As the temperature of a blackbody increases, the intensity of the radiation increases, while the peak of the distribution shifts to shorter wavelengths.

$$\lambda_{max}T = 2.898 \times 10^{-3} \text{ m} \cdot \text{K} \tag{28.1}$$

Vibrating molecules are characterized by discrete energy levels called quantum states. Each quantum state is represented by a quantum number, n.

$$E_n = nhf \tag{28.2}$$

$$h = 6.626 \times 10^{-34} \text{ J} \cdot \text{s}$$

Molecules emit or absorb energy in discrete units of light energy called quanta. The energy of a quantum or photon corresponds to the energy difference between adjacent quantum states.

$$E = hf \qquad (28.3)$$

When light is incident on certain metallic surfaces, electrons can be emitted from the surfaces. This is the photoelectric effect, discovered by Hertz. One cannot explain many of the features of the photoelectric effect using classical concepts. However, in 1905, Einstein provided a successful explanation of the effect by extending Planck's quantum concept to include electromagnetic fields. In his model, Einstein assumed that light consists of a stream of particles called **photons** whose energy is given by $E = hf$, where h is Planck's constant and f is their frequency.

The maximum kinetic energy of a photoelectron also depends on the work function of the metal, ϕ, which is typically a few eV. This model is in excellent agreement with experimental results, including the prediction of a cutoff (or threshold) wavelength above which no photoelectric effect is observed. The work function represents the minimum energy with which an electron is bound in a metal.

$$K_{max} = hf - \phi \qquad (28.5)$$

$$\lambda_c = \frac{hc}{\phi} \qquad (28.7)$$

The Compton effect involves the scattering of an x-ray by an electron. The scattered x-ray undergoes a change in wavelength called the Compton shift, which cannot be explained using classical concepts.

$$\lambda' - \lambda_0 = \frac{h}{m_e c}(1 - \cos\theta) \qquad (28.8)$$

By treating the x-ray as a photon (the quantum concept), the scattering process between the photon and electron predicts a shift in photon (x-ray) wavelength, where θ is the angle between the incident and scattered x-ray and m_e is the mass of the electron. The formula is in excellent agreement with experimental results.

The quantity $\dfrac{h}{m_e c} = 0.00243$ nm is called the Compton wavelength.

According to the de Broglie hypothesis, a material particle should have an associated wavelength λ, which depends on its momentum.

$$\lambda = \frac{h}{mv} \qquad (28.10)$$

Matter waves can be represented by a wave function Ψ, which, in general, depends on position and time. In the expression for the part ψ of the wave function which depends only on position, and which describes a freely moving particle, $k = 2\pi/\lambda$ is the wave number.

$$\psi(x) = A\sin\left(\frac{2\pi x}{\lambda}\right) = A\sin(kx) \qquad (28.20)$$

The normalization condition is a statement of the requirement that the particle exists at some point (along the x axis in the one-dimensional case) at all times.

$$\int_{-\infty}^{\infty} |\psi|^2 \, dx = 1 \tag{28.22}$$

Although it is not possible to specify the position of a particle exactly, it is possible to calculate the probability P_{ab} of finding the particle within an interval $a \leq x \leq b$.

$$P_{ab} = \int_a^b |\psi|^2 \, dx \tag{28.23}$$

The average of many measured values for the position of a particle is called the expectation value of the coordinate x.

$$\langle x \rangle \equiv \int_{-\infty}^{\infty} |\psi|^2 x \, dx \tag{28.24}$$

The uncertainty principle states that if a measurement of position is made with a precision Δx and a **simultaneous** measurement of momentum is made with a precision Δp, then the product of the two uncertainties can never be smaller than $\hbar/2$.

$$\Delta x \Delta p_x \geq \frac{\hbar}{2} = \frac{h}{4\pi} \tag{28.17}$$

The allowed wave functions for a particle in a rigid box of width L are sinusoidal.

$$\psi(x) = A \sin\left(\frac{n\pi x}{L}\right) \tag{28.28}$$

$$n = 1, 2, 3, \ldots$$

The energy of a particle in a box is quantized and the least energy which the particle can have is called the zero-point energy.

$$E_n = \left(\frac{h^2}{8mL^2}\right) n^2 \tag{28.29}$$

$$n = 1, 2, 3, \ldots$$

The time independent Schrödinger equation for a particle confined to moving along the x-axis (total energy E is constant) allows in principle the determination of the wave functions and energies of the allowed states if the potential energy function is known.

$$\frac{-\hbar^2}{2m}\frac{d^2\psi}{dx^2} + U\psi = E\psi \qquad (28.30)$$

REVIEW CHECKLIST

▷ Describe the account of blackbody radiation proposed by Planck.

▷ Describe the Einstein model for the photoelectric effect, and the predictions of the fundamental photoelectric effect equation for the maximum kinetic energy of photoelectrons. Recognize that Einstein's model of the photoelectric effect involves the photon concept ($E = hf$), and that the basic features of the photoelectric effect are consistent with this model.

▷ Describe the Compton effect (the scattering of x-rays by electrons) and be able to use the formula for the Compton shift. Recognize that the Compton effect can only be explained using the photon concept.

▷ Discuss the wave properties of particles, the de Broglie wavelength concept, and the dual nature of both matter and light.

▷ Describe the concept of wave function for the representation of matter waves and state in equation form the normalization condition and expectation value of the coordinate.

▷ Discuss the manner in which the uncertainty principle makes possible a better understanding of the dual wave-particle nature of light and matter.

ANSWERS TO SELECTED CONCEPTUAL QUESTIONS

2. If matter has a wave nature, why is this wave-like characteristic not observable in our daily experience?

Answer. For any object that we can perceive directly, then de Broglie wavelength $\lambda = h/mv$ is too small to be measured by any means; therefore, no wavelike characteristics can be observed. The object will not diffract noticeably when it goes through an aperture. It will not show resolvable interference maxima and minima when it goes through two openings. It will not show resolvable nodes and antinodes if it is in resonance.

□   □   □   □

6. Why does the existence of a cutoff frequency in the photoelectric effect favor a particle theory of light rather than a wave theory?

Answer Wave theory predicts that the photoelectric effect should occur at any frequency, provided that the light intensity is high enough. As is implied by the question, this is in contradiction to experimental results.

□   □   □   □

12. An x-ray photon is scattered by an electron. What happens to the frequency of the scattered photon relative to that of the incident photon?

Answer The x-ray photon transfers some of its energy to the electron. Thus, its energy, and therefore its frequency, must be decreased.

□   □   □   □

18. If the photoelectric effect is observed for one metal, can you conclude that the effect will also be observed for another metal under the same conditions? Explain.

Answer No. Suppose that the incident light frequency at which you first observed the photoelectric effect is above the cutoff frequency of the first metal, but less than the cutoff frequency of the second metal. In that case, the photoelectric effect would not be observed at all in the second metal.

SOLUTIONS TO SELECTED END-OF-CHAPTER PROBLEMS

5. An FM radio transmitter has a power output of 150 kW and operates at a frequency of 99.7 MHz. How many photons per second does the transmitter emit?

Solution

Each photon has an energy

$$E = hf = (6.63 \times 10^{-34} \text{ J} \cdot \text{s})(99.7 \times 10^{-6} \text{ s}^{-1}) = 6.61 \times 10^{-26} \text{ J}$$

The number of photons per second is the power divided by the energy per photon:

$$R = \frac{\mathcal{P}}{E} = \frac{150 \times 10^3 \text{ J/s}}{6.61 \times 10^{-26} \text{ J}} = 2.27 \times 10^{30} \text{ photons/s} \qquad \Diamond$$

9. Two light sources are used in a photoelectric experiment to determine the work function for a particular metal surface. When green light from a mercury lamp ($\lambda = 546.1$ nm) is used, a retarding potential of 0.376 V reduces the photocurrent to zero. (a) Based on this measurement, what is the work function for this metal? (b) What stopping potential is observed when using the yellow light from a helium discharge tube ($\lambda = 587.5$ nm)?

Solution

According to Table 28.1, the work function for most metals is on the order of a few eV, so this metal is probably similar. We can expect the stopping potential for the yellow light to be slightly lower than 0.376 V since the yellow light has a longer wavelength (lower frequency) and therefore less energy than the green light.

In this photoelectric experiment, the green light has sufficient energy hf to overcome the work function of the metal ϕ so that the ejected electrons have a maximum kinetic energy of 0.376 eV. With this information, we can use the photoelectric effect equation to find the work function, which can then be used to find the stopping potential for the less energetic yellow light.

(a) Einstein's photoelectric effect equation is $K_{max} = hf - \phi$, and the energy required to raise an electron through a 1-V potential is 1 eV, so that $K_{max} = e\Delta V_s = 0.376$ eV.

The energy of a photon from the mercury lamp is:

$$hf = \frac{hc}{\lambda} = \frac{\left(4.14 \times 10^{-15} \text{ eV} \cdot \text{s}\right)\left(3.00 \times 10^8 \text{ m/s}\right)}{546.1 \times 10^{-9} \text{ m}} = 2.27 \text{ eV}$$

Therefore, the work function for this metal is:

$$\phi = hf - K_{max} = 2.27 \text{ eV} - 0.376 \text{ eV} = 1.90 \text{ eV} \qquad \Diamond$$

(b) For the yellow light, $\lambda = 587.5$ nm

and $\qquad hf = \frac{hc}{\lambda} = \frac{\left(4.14 \times 10^{-15} \text{ eV} \cdot \text{s}\right)\left(3.00 \times 10^8 \text{ m/s}\right)}{587.5 \times 10^{-9} \text{ m}} = 2.11 \text{ eV}$

Therefore, $\qquad K_{max} = hf - \phi = 2.11 \text{ eV} - 1.89 \text{ eV} = 0.216 \text{ eV}$

so $\qquad \Delta V_s = 0.216 \text{ V} \qquad \Diamond$

The work function for this metal is lower than we expected, and does not correspond with any of the values in Table 28.1. Further examination in the **CRC Handbook of Chemistry and Physics** reveals that all of the metal elements have work functions between 2 and 6 eV. However, a single metal's work function may vary by about 1 eV depending on impurities in the metal, so it is just barely possible that a metal might have a work function of 1.90 eV.

The stopping potential for the yellow light is indeed lower than for the green light as we expected. An interesting calculation is to find the wavelength for the lowest energy light that will eject electrons from this metal. That threshold wavelength for K_{max} is 658 nm, which is red light in the visible portion of the electromagnetic spectrum.

15. A 0.00160-nm photon scatters from a free electron. For what (photon) scattering angle does the recoiling electron have kinetic energy equal to the energy of the scattered photon?

Solution The energy of the incoming photon is

$$E_0 = \frac{hc}{\lambda} = \frac{\left(6.63 \times 10^{-34}\ \text{J} \cdot \text{s}\right)\left(3.00 \times 10^8\ \text{m/s}\right)}{0.00160 \times 10^{-9}\ \text{m}} = 1.24 \times 10^{-13}\ \text{J}$$

Since the outgoing photon and the electron each have half of this energy in kinetic form,

$$E' = 6.22 \times 10^{-14}\ \text{J} \qquad \text{and} \qquad \lambda' = \frac{hc}{E'} = 3.20 \times 10^{-12}\ \text{m}$$

The shift in wavelength is

$$\Delta\lambda = \lambda' \lambda\!\!\!\!- \quad = 1.60 \times 10^{-12}\ \text{m}$$

But by Equation 28.8,

$$\Delta\lambda = \lambda_c (1 - \cos\theta)$$

so

$$\cos\theta = 1 - \frac{\Delta\lambda}{\lambda_c} = 1 - \frac{1.60 \times 10^{-12}\ \text{m}}{0.00243 \times 10^{-9}\ \text{m}} = 0.342$$

and

$$\theta = 70.0°$$ ◊

---

21. The nucleus of an atom is on the order of 10^{-14} m in diameter. For an electron to be confined to a nucleus, its de Broglie wavelength would have to be of this order of magnitude or smaller. (a) What would be the kinetic energy of an electron confined to this region? (b) On the basis of this result, would you expect to find an electron in a nucleus? Explain.

Solution The de Broglie wavelength of a normal ground-state orbiting electron is on the order 10^{-10} m (the diameter of a hydrogen atom) so with a shorter wavelength, the electron would have more kinetic energy if confined inside the nucleus. If the kinetic energy is much greater than the potential energy characterizing its attraction with the positive nucleus, then the electron will escape from its electrostatic potential well.

If we try to calculate the velocity of the electron from the de Broglie wavelength, we find that

$$v = \frac{h}{m_e \lambda} = \frac{6.63 \times 10^{-34}\ \text{J} \cdot \text{s}}{\left(9.11 \times 10^{-31}\ \text{kg}\right)\left(10^{-14}\ \text{m}\right)} = 7.27 \times 10^{10}\ \text{m/s}$$

which is not possible since it exceeds the speed of light. Therefore, we must use the relativistic energy expression to find the kinetic energy of this fast-moving electron.

(a)   We find the momentum of the particle:

$$p = \frac{h}{\lambda} = \frac{6.63 \times 10^{-34} \text{ J} \cdot \text{s}}{10^{-14} \text{ m}} = 6.63 \times 10^{-20} \text{ N} \cdot \text{s}$$

We find the particle's relativistic total energy from $E^2 = (pc)^2 + (mc^2)^2$:

$$E = \sqrt{\left(1.99 \times 10^{-11} \text{ J}\right)^2 + \left(8.19 \times 10^{-14} \text{ J}\right)^2} = 1.99 \times 10^{-11} \text{ J}$$

Its relativistic kinetic energy is $K = E - mc^2$:

$$K = \left(1.99 \times 10^{-11} \text{ J} - 8.19 \times 10^{-14} \text{ J}\right)\left(\frac{1}{1.60 \times 10^{-19} \text{ J/eV}}\right) = 124 \text{ MeV} \sim 100 \text{ MeV} \quad \Diamond$$

(b)   The electrostatic potential energy of an electron-proton system with a separation of 10^{-14} m is

$$U = -\frac{k_e e^2}{r} = -\frac{\left(8.99 \times 10^9 \text{ N} \cdot \text{m}^2 / \text{C}^2\right)\left(1.60 \times 10^{-19} \text{ C}\right)^2}{10^{-14} \text{ m}} = -2.30 \times 10^{-14} \text{ J} \sim -0.1 \text{ eV} \quad \Diamond$$

Since the kinetic energy is nearly 1000 times greater than the potential energy, the electron would immediately escape the proton's attraction and would not be confined to the nucleus. $\quad \Diamond$

It is also interesting to notice in the above calculations that the rest energy of the electron is negligible compared to the momentum contribution to the total energy.

---

25.   Neutrons traveling at 0.400 m/s are directed through a double slit having a 1.00-mm separation. An array of detectors is placed 10.0 m from the slit. (a) What is the de Broglie wavelength of the neutrons? (b) How far off axis is the first zero-intensity point on the detector array? (c) When a neutron reaches a detector, can we say which slit the neutron passed through? Explain.

Solution   We use the waves in interference model.

(a) $\lambda = \dfrac{h}{mv} = \dfrac{6.63 \times 10^{-34} \text{ J} \cdot \text{s}}{\left(1.67 \times 10^{-27} \text{ kg}\right)(0.400 \text{ m/s})} = 9.93 \times 10^{-7} \text{ m}$ $\quad \Diamond$

(b) The condition for destructive interference in a multiple-slit experiment is

$$d \sin\theta = \left(m + \frac{1}{2}\right)\lambda \text{ with } m = 0 \text{ for the first minimum.}$$

Then,

$$\theta = \sin^{-1}\left(\frac{\lambda}{2d}\right) = 0.0284°$$

$\dfrac{y}{L} = \tan\theta$:

$$y = L\tan\theta = (10.0 \text{ m})\tan(0.0284°) = 4.96 \text{ mm}$$ ◊

(c) We cannot say the neutron passed through one slit. We can only say it passed through the pair of slits, as a water wave does to produce an interference pattern.   ◊

---

27. An electron ($m_e = 9.11 \times 10^{-31}$ kg) and a bullet ($m = 0.0200$ kg) each have a velocity of magnitude 500 m/s, accurate to within 0.0100%. Within what limits could we determine the position of the objects along the direction of the velocity?

Solution   We use the quantum particle model for both objects. For the electron, the uncertainty in momentum is

$$\Delta p = m_e \Delta v = \left(9.11 \times 10^{-31} \text{ kg}\right)(500 \text{ m/s})\left(1.00 \times 10^{-4}\right) = 4.56 \times 10^{-32} \text{ kg} \cdot \text{m/s}$$

The minimum uncertainty in position is then

$$\Delta x = \frac{h}{4\pi \Delta p} = \frac{6.63 \times 10^{-34} \text{ J} \cdot \text{s}}{4\pi\left(4.56 \times 10^{-32} \text{ kg} \cdot \text{m/s}\right)} = 1.16 \text{ mm}$$ ◊

For the bullet, $\Delta p = m\Delta v = (0.0200 \text{ kg})(500 \text{ m/s})\left(1.00 \times 10^{-4}\right) = 1.00 \times 10^{-3} \text{ kg} \cdot \text{m/s}$

$$\Delta x = \frac{h}{4\pi \Delta p} = 5.28 \times 10^{-32} \text{ m}$$ ◊

Quantum mechanics describes all objects, but the quantum fuzziness in position is unobservably small for the bullet and large for the small-mass electron.

---

33. A free electron has a wave function $\psi(x) = A \sin(5.00 \times 10^{10} x)$ where x is in meters. Find (a) the de Broglie wavelength, (b) the linear momentum, and (c) the kinetic energy in electron volts.

Solution

(a) The wave function $\psi(x) = A \sin\left(5.00 \times 10^{10} x\right)$ will go through one full cycle between $x_1 = 0$ and $\left(5.00 \times 10^{10}\right)x_2 = 2\pi$. The wavelength is then

$$\lambda = x_2 - x_1 = \frac{2\pi}{5.00 \times 10^{10} \text{ m}^{-1}} = 1.26 \times 10^{-10} \text{ m} \qquad \Diamond$$

To say the same thing, we can inspect $A \sin\left[\left(5.00 \times 10^{10} \text{ m}^{-1}\right)x\right]$ to see that the wave number is $k = 5.00 \times 10^{10} \text{ m}^{-1}$.

(b) Since $\lambda = \dfrac{h}{p}$, the momentum is

$$p = \frac{h}{\lambda} = \frac{6.63 \times 10^{-34} \text{ J} \cdot \text{s}}{1.26 \times 10^{-10} \text{ m}} = 5.27 \times 10^{-24} \text{ kg} \cdot \text{m/s} \qquad \Diamond$$

(c) The electron's kinetic energy is

$$K = \frac{1}{2}mv^2 = \frac{p^2}{2m} = \frac{5.27 \times 10^{-24} \text{ kg} \cdot \text{m/s}}{2\left(9.11 \times 10^{-31} \text{ kg}\right)}\left(\frac{1 \text{ eV}}{1.60 \times 10^{-19} \text{ J}}\right) = 95.5 \text{ eV} \qquad \Diamond$$

Its relativistic total energy is $511 \text{ keV} + 95.5 \text{ eV}$.

---

35. An electron is contained in a one-dimensional box of width 0.100 nm. (a) Draw an energy level diagram for the electron for levels up to $n = 4$. (b) Find the wavelengths of all photons that can be emitted by the electron in making spontaneous transitions that could eventually carry it from the $n = 4$ to the $n = 1$ state.

Solution

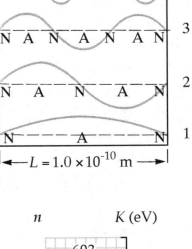

(a) We can draw a diagram that parallels our treatment of mechanical waves under boundary conditions. In each standing-wave state, we measure the distance d from one node to another (N to N), and base our solution upon that:

Since $\qquad d_{\text{N to N}} = \dfrac{\lambda}{2}$ and $\lambda = \dfrac{h}{p}$, $\qquad p = \dfrac{h}{\lambda} = \dfrac{h}{2d}$

Next, $\qquad K = \dfrac{p^2}{2m_e} = \dfrac{h^2}{8m_e d^2} = \dfrac{1}{d^2}\left(\dfrac{\left(6.63 \times 10^{-34}\ \text{J}\cdot\text{s}\right)^2}{8\left(9.11 \times 10^{-31}\ \text{kg}\right)}\right)$

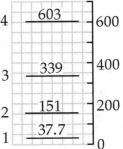

Evaluating, $\quad K = \dfrac{6.03 \times 10^{-38}\ \text{J}\cdot\text{m}^2}{d^2} = \dfrac{3.77 \times 10^{-19}\ \text{eV}\cdot\text{m}^2}{d^2}$

In state 1, $\quad d = 1.00 \times 10^{-10}$ m $\qquad K_1 = 37.7$ eV

In state 2, $\quad d = 5.00 \times 10^{-11}$ m $\qquad K_2 = 151$ eV

In state 3, $\quad d = 3.33 \times 10^{-11}$ m $\qquad K_3 = 339$ eV

In state 4, $\quad d = 2.50 \times 10^{-11}$ m $\qquad K_4 = 603$ eV

These energy levels are shown in the diagram to the right. $\qquad\qquad\qquad\quad \lozenge$

(b) When the charged, massive electron inside the box makes a downward transition from one energy level to another, a chargeless, massless photon comes out of the box, carrying the difference in energy, ΔE. Its wavelength is

$$\lambda = \frac{c}{f} = \frac{hc}{\Delta E} = \frac{\left(6.63 \times 10^{-34}\ \text{J}\cdot\text{s}\right)\left(3.00 \times 10^8\ \text{m/s}\right)}{\Delta E\left(1.602 \times 10^{-19}\ \text{J/eV}\right)} = \frac{1.24 \times 10^{-6}\ \text{eV}\cdot\text{m}}{\Delta E}$$

Transition	$4 \to 3$	$4 \to 2$	$4 \to 1$	$3 \to 2$	$3 \to 1$	$2 \to 1$
ΔE (eV)	264	452	565	188	302	113
Wavelength (nm)	4.71	2.75	2.20	6.60	4.12	11.0

The wavelengths of light released for each transition are given in the table above. $\qquad \lozenge$

41. Show that the wave function $\psi = A e^{i(kx-\omega t)}$ is a solution to the Schrödinger equation (See Equation 28.30), where $k = 2\pi/\lambda$ and $U = 0$.

Solution   From

$$\psi = A e^{i(kx-\omega t)} \qquad (1)$$

we evaluate

$$\frac{d\psi}{dx} = ik A e^{i(kx-\omega t)}$$

and

$$\frac{d^2\psi}{dx^2} = -k^2 A e^{i(kx-\omega t)} \qquad (2)$$

We substitute Equations (1) and (2) into the Schrödinger equation, so that

$$\frac{d^2\psi}{dx^2} = -\frac{2m}{\hbar^2}(E-U)\psi \qquad \text{(Eq. 28.30)}$$

becomes

$$-k^2 A e^{i(kx-\omega t)} = -\left(\frac{2m}{\hbar^2}K\right) A e^{i(kx-\omega t)} \qquad (3)$$

The wave function $\psi = A e^{i(kx-\omega t)}$ is a solution to the Schrödinger equation if Equation (3) is true. Both sides depend on A, x, and t in the same way, so we can cancel several terms, and determine that we have a solution if:

$$k^2 = \frac{2m}{\hbar^2}K$$

But this is true for a nonrelativistic particle with mass,

since

$$\frac{2m}{\hbar^2}K = \frac{2m}{(h/2\pi)^2}\left(\tfrac{1}{2}mv^2\right) = \frac{4\pi^2 m^2 v^2}{h^2} = \left(\frac{2\pi p}{h}\right)^2 = \left(\frac{2\pi}{\lambda}\right)^2 = k^2$$

Therefore, the given wave function does satisfy Equation 28.30.   ◊

---

45. An electron with kinetic energy $E = 5.00$ eV is incident on a barrier with thickness $L = 0.200$ nm and height $U = 10.0$ eV (See Fig. P28.44). What is the probability that the electron (a) will tunnel through the barrier? (b) will be reflected?

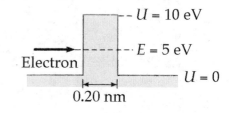

Figure P28.44

Solution We are using the quantum particle under boundary conditions model. The transmission coefficient is given by Equation 28.36, $T \cong e^{-2CL}$, where C is the decay constant for the wave function inside the barrier:

$$C = \frac{\sqrt{2m(U-E)}}{\hbar} = \frac{\sqrt{2(9.11 \times 10^{-31}\ \text{kg})(10.0\ \text{eV} - 5.00\ \text{eV})(1.60 \times 10^{-19}\ \text{J/eV})}}{6.63 \times 10^{-34}\ \text{J} \cdot \text{s}/2\pi} = 1.14 \times 10^{10}\ \text{m}^{-1}$$

(a) For transmission, $T \cong e^{-2(1.14 \times 10^{10}\ \text{m}^{-1})(2.00 \times 10^{-10}\ \text{m})} = e^{-4.58} = 0.0103$ ◊

(b) If the electron does not tunnel, it is reflected, with probability $1 - 0.010 = 0.990$ ◊

Related Commentary A typical scanning – tunneling electron microscope (STM) can be built for less than $5000, and can be tied directly to a home computer.

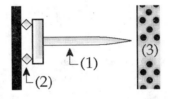

The STM essentially consists of a needle (1) that is mounted on a few piezo-electric crystals (2). When a voltage is induced across the piezo-electric crystals, the crystals change their shape, and the needle moves.

The STM charges the needle, and moves the tip of the needle to within a few tenths of a nanometer of the test sample (3); the remaining gap provides the energy barrier that is required for tunneling. When the electron cloud of one of the needle's atoms is close to the electron cloud of one of the sample's atoms, a relatively large number of electrons tunnels from one to the other, generating a current. The current is then measured by a computer. As the needle moves across the surface of the sample, an image of the atoms is generated.

---

51. The following table shows data obtained in a photoelectric experiment. (a) Using these data, make a graph similar to Figure 28.8 that plots as a straight line. From the graph, determine (b) an experimental value for Planck's constant (in joule-seconds) and (c) the work function (in electron volts) for the surface. (Two significant figures for each answer are sufficient.)

Wavelength (nm)	Maximum Kinetic Energy of Photoelectrons (eV)
588	0.67
505	0.98
445	1.35
399	1.63

Solution Convert each wavelength to a frequency using the relation $\lambda f = c$, where c is the speed of light:

$\lambda_1 = 588 \times 10^{-9}$ m $\qquad f_1 = 5.10 \times 10^{14}$ Hz

$\lambda_2 = 505 \times 10^{-9}$ m $\qquad f_2 = 5.94 \times 10^{14}$ Hz

$\lambda_3 = 445 \times 10^{-9}$ m $\qquad f_3 = 6.74 \times 10^{14}$ Hz

$\lambda_4 = 399 \times 10^{-9}$ m $\qquad f_4 = 7.52 \times 10^{14}$ Hz

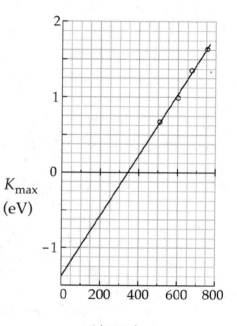

K_{max} (eV)

f (THz)

(a) Plot each point on an energy vs. frequency graph, as shown at the right. Extend a line through the set of 4 points, as far as the $-y$ intercept.

(b) Our basic equation is $K_{max} = hf - \phi$. Therefore, Planck's constant should be equal to the slope of the K-f graph, which can be found from a least-squares fit or from reading the graph as:

$$h_{exp} = \frac{\text{Rise}}{\text{Run}} = \frac{1.25 \text{ eV} - 0.25 \text{ eV}}{6.5 \times 10^{14} \text{ Hz} - 4.0 \times 10^{14} \text{ Hz}} = 4.0 \times 10^{-15} \text{ eV} \cdot \text{s} = 6.4 \times 10^{-34} \text{ J} \cdot \text{s} \qquad \lozenge$$

From the scatter on the graph, we estimate the uncertainty as 3%.

(c) From the same equation, the work function for the surface is the negative of the y-intercept of the graph. You can extend the graph until the line intercepts the y axis, and then check your accuracy with a calculation similar to the one below:

$$\phi = -\left(K_{max} - h_{exp}f\right) = -\left(0.67 \text{ eV} - \left(4.0 \times 10^{-15} \text{ eV} \cdot \text{s}\right)\left(5.10 \times 10^{14} \text{ s}^{-1}\right)\right) = 1.4 \text{ eV} \qquad \lozenge$$

55. An electron is represented by the time-independent wave function

$$\psi(x) = \begin{cases} Ae^{-\alpha x} & \text{for } x > 0 \\ Ae^{+\alpha x} & \text{for } x < 0 \end{cases}$$

(a) Sketch the wave function as a function of x. (b) Sketch the probability that the electron is found between x and $x + dx$. (c) Argue that this can be a physically reasonable wave function. (d) Normalize the wave function. (e) Determine the probability of finding the electron somewhere in the range

$$x_1 = -\frac{1}{2\alpha} \qquad \text{to} \qquad x_2 = \frac{1}{2\alpha}.$$

Solution

(a)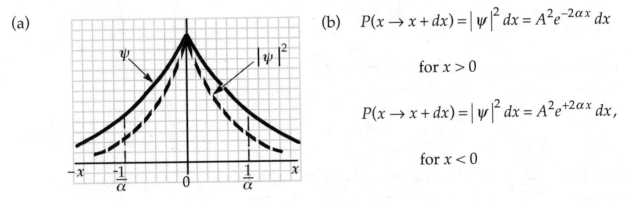

(b) $P(x \to x + dx) = |\psi|^2 dx = A^2 e^{-2\alpha x} dx$

for $x > 0$

$P(x \to x + dx) = |\psi|^2 dx = A^2 e^{+2\alpha x} dx,$

for $x < 0$

(c) ψ is continuous; $\psi \to 0$ as $x \to \infty$; it mimics an electron bound at $x = 0$. ◊

(d) Since ψ is symmetric,

$$\int_{-\infty}^{\infty} |\psi|^2 dx = 2 \int_0^{\infty} |\psi|^2 dx = 1$$

or

$$2A^2 \int_0^{\infty} e^{-2\alpha x} dx = 1$$

or

$$\left(\frac{2A^2}{-2\alpha}\right)\left(e^{-\infty} - e^0\right) = 1$$

This gives

$$A = \sqrt{\alpha} \qquad ◊$$

(e) $P_{(-1/2\alpha) \to (1/2\alpha)} = 2\left(\sqrt{\alpha}\right)^2 e^{-2\alpha x} dx = \left(\frac{2\alpha}{-2\alpha}\right)\left(e^{-2\alpha/2\alpha} - 1\right) = 1 - e^{-1} = 0.632$ ◊

Chapter 29

Atomic Physics

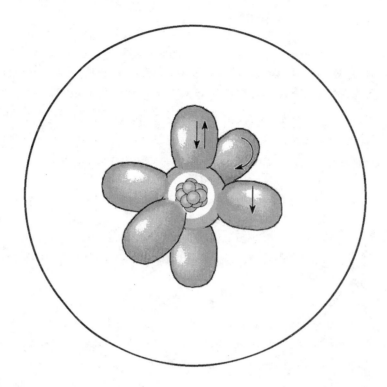

INTRODUCTION

A large portion of this chapter is an application of quantum physics to the study of the hydrogen atom. Understanding the hydrogen atom, the simplest atomic system, is especially important for several reasons:

- Much of what is learned about the hydrogen atom with its single electron can be extended to such single-electron ions as He^+ and Li^{2+}.

- The hydrogen atom is an ideal system for performing precise tests of theory against experiment and for improving our overall understanding of atomic structure.

- The quantum numbers used to characterize the allowed states of hydrogen can be used to describe the allowed states of more complex atoms. This enables us to understand the periodic table of the elements, one of the greatest triumphs of quantum physics.

- The basic ideas about electrons in atoms provide background for studying nuclei.

NOTES FROM SELECTED CHAPTER SECTIONS

29.1 Early Structural Models of the Atom

A need for modification of the Bohr theory became apparent when improved spectroscopic techniques were used to examine the spectral lines of hydrogen. It was found that many of the lines in the Balmer and other series were not single lines at all. Instead, each was a group of lines spaced very close together. An additional difficulty arose when it was observed that, in some situations, certain single spectral lines were split into three closely spaced lines when the atoms were placed in a strong magnetic field.

29.2 The Hydrogen Atom Revisited

In the three-dimensional problem of the hydrogen atom, three quantum numbers are required for each stationary state, corresponding to the three independent degrees of freedom for the electron.

The three quantum numbers which emerge from the theory are represented by the symbols n, ℓ, and m_ℓ. The quantum number n is called the **principal quantum number**; ℓ is called the **orbital quantum number**; and m_ℓ is called the **orbital magnetic quantum number.**

There are certain important relationships between these quantum numbers, as well as certain restrictions on their values. These restrictions are:

The three quantum numbers are integers.
The values of n can range from 1 to ∞.
The values of ℓ can range from 0 to $n-1$.
The values of m_ℓ can range from $-\ell$ to ℓ.

For historical reasons, **all states with the same principal quantum number are said to form a shell.** These shells are identified by the letters K, L, M, . . . , which designate the states for which $n = 1, 2, 3, \ldots$. Likewise, **the states having the same values of n and ℓ are said to form a subshell.** The letters $s, p, d, f, g, h, \ldots$ are used to designate the states for which $\ell = 0, 1, 2, 3, \ldots$.

29.3 The Spin Magnetic Quantum Number

The **spin magnetic quantum number** m_s accounts for the two closely spaced energy states corresponding to the two possible orientations of the electron spin.

29.4 The Wave Functions for Hydrogen

Since the potential energy function for the hydrogen atom depends only on the radial distance r, the wave functions that describe the s states are **spherically symmetric.**

29.5 Physical Interpretation of the Quantum Numbers

The allowed energy levels for the electron in a hydrogen atom are determined by the value of the **principal quantum number,** n. There are three "other" quantum numbers, which also serve to characterize the wave function of the hydrogen electron. The **orbital quantum number,** ℓ, restricts the orbital angular momentum to discrete values. The **magnetic orbital quantum number,** m_ℓ, specifies the allowed directions of the angular momentum with respect to an external magnetic field. The **spin magnetic quantum number,** m_s, restricts the spin angular momentum of the hydrogen electron to have only two possible orientations in space. The spin contribution to the magnetic moment is twice the contribution of the orbital motion.

29.6 The Exclusion Principle and the Periodic Table

The **Pauli exclusion principle** states that no two electrons can exist in identical quantum states. This means that no two electrons in a given atom can be characterized by the same set of quantum numbers at the same time.

Hund's rule states that when an atom has orbitals of equal energy, the order in which they are filled by electrons is such that a maximum number of electrons will have unpaired spins.

29.7 Atomic Spectra: Visible and X-Ray

An atom will emit electromagnetic radiation if the atom in an excited state makes a transition to a lower energy state. The set of wavelengths observed for a species by such processes is called an **emission spectrum**. Likewise, atoms in the ground-state configuration can also absorb electromagnetic radiation at specific wavelengths, giving rise to an **absorption spectrum.** Such spectra can be used to identify the elements in gases.

Since the orbital angular momentum of an atom changes when a photon is emitted or absorbed (that is, as a result of a transition) and since angular momentum must be conserved, we conclude that **the photon involved in the process must carry angular momentum.**

X-rays are emitted by atoms when an electron undergoes a transition from an outer shell into an electron vacancy in one of the inner shells. Transitions into a vacant state in the K shell give rise to the K series of spectral lines; in a similar way, transitions into a vacant state in the L shell create the L series of lines, and so on. The x-ray spectrum of a metal target consists of a set of sharp characteristic lines superimposed on a broad, continuous spectrum.

EQUATIONS AND CONCEPTS

The **potential energy** of the hydrogen atom depends on the distance of the electron from the nucleus.

$$U(r) = -k_e \frac{e^2}{r} \tag{29.1}$$

The energy states allowed in the Bohr model for the hydrogen atom depend only on the quantum number, n, the **principal quantum number**.

$$E_n = -\frac{13.606}{n^2} \text{ eV} \tag{29.2}$$

$$(n = 1, 2, 3 \ldots)$$

The lowest energy state or ground state corresponds to the principal quantum number $n = 1$. The energy level approaches $E = 0$ as n approaches infinity. This is the **ionization energy** for the atom.

The simplest wave function for hydrogen is the one that describes the $1s$ state and depends only on the radial distance r. The parameter a_0 is the Bohr radius.

$$\psi_{1s}(r) = \frac{1}{\sqrt{\pi a_0^3}} e^{-r/a_0} \tag{29.3}$$

$$a_0 = \frac{\hbar^2}{m_e k_e e^2} = 0.0529 \text{ nm}$$

The radial probability density for the 1s state of hydrogen is defined as the probability per unit radial distance of finding the electron in a spherical shell of radius r and thickness dr.

$$P_{1s}(r) = \left(\frac{4r^2}{a_0^3}\right) e^{-2r/a_o} \qquad (29.7)$$

The value of the orbital quantum number ℓ determines the magnitude of the electron's **orbital angular momentum,** L.

$$|\mathbf{L}| = \sqrt{\ell(\ell+1)}\,\hbar \qquad (29.9)$$

$$(\ell = 0,\ 1,\ 2,\ 3,\ \ldots,\ n-1)$$

The magnetic orbital quantum number, m_ℓ, specifies the allowed values of the component of $\mathbf{L}$ along the direction of a z-axis. This effect is referred to as **space quantization**.

$$L_z = m_\ell \hbar \qquad (29.10)$$

$$(m_\ell = -\ell,\ -\ell+1,\ -\ell+2,\ \ldots,\ \ell)$$

In a weak external magnetic field, the angle, θ, between the direction of $\mathbf{L}$ for an orbiting electron and the direction of the z-axis is quantized. L can never be parallel or antiparallel to the z-axis (θ cannot be zero).

$$\cos\theta = \frac{L_z}{|\mathbf{L}|} = \frac{m_\ell}{\sqrt{\ell(\ell+1)}} \qquad (29.11)$$

In addition to orbital angular momentum, the electron has an intrinsic angular momentum or **spin angular momentum** S, as if due to spinning on its axis. This spin angular momentum is described by a single quantum number s whose value can only be $\frac{1}{2}$.

$$S = \sqrt{s(s+1)}\,\hbar = \frac{\sqrt{3}}{2}\hbar \qquad (29.12)$$

The spin angular momentum is space quantized with respect to the direction of an external magnetic field (along the z direction). The spin magnetic quantum number m_s can have values of $\pm\frac{1}{2}$.

$$S_z = m_s\hbar = \pm\frac{1}{2}\hbar \qquad (29.13)$$

The **spin magnetic moment** μ_s is related to the spin angular momentum.

$$\mu_s = \left(-\frac{e}{m_e}\right)\mathbf{S} \qquad (29.14)$$

Atomic transitions between two energy levels are governed by **selection rules**.

$$\Delta \ell = \pm 1 \quad \text{and} \quad \Delta m_\ell = 0 \text{ or } \pm 1 \qquad (29.16)$$

The shielding effect of the nuclear charge by inner-core electrons must be taken into account when calculating the allowed energy levels of multiple-electron atoms. The atomic number is replaced by an **effective atomic number**, Z_{eff}, which depends on the values of n and ℓ.

$$E_n = -\frac{(13.606 \text{ eV})Z_{eff}^2}{n^2} \qquad (29.18)$$

SUGGESTIONS, SKILLS, AND STRATEGIES

Before reading this chapter, it may be helpful to review Section 5 of Chapter 11. This section deals with the Bohr model of the atom, and makes it easier to understand the modern concept of the atomic structure.

After reading the chapter in your text, review the significance of each of the quantum numbers that is used to describe the various electronic states of electrons in an atom. Next, review the set of allowed values for each of the quantum numbers.

In addition to the principal quantum number n (which can range from 1 to ∞), other quantum numbers are necessary to specify completely the possible energy levels in the hydrogen atom and also in more complex atoms.

All energy states with the same principal quantum number, n, form a shell. These shells are identified by the spectroscopic notation K, L, M, ... corresponding to $n = 1, 2, 3, \ldots$.

The orbital quantum number ℓ [which can range from 0 to $(n - 1)$], determines the allowed value of orbital angular momentum. All energy states having the same values of n and ℓ form a subshell. The letter designations $s, p, d, f, \ldots$ correspond to values of $\ell = 0, 1, 2, 3, \ldots$.

The magnetic orbital quantum number m_ℓ (which can range from $-\ell$ to ℓ) determines the possible orientations of the electron's orbital angular momentum vector in the presence of an external magnetic field.

The spin magnetic quantum number m_s, can have only two values, $m_s = -\frac{1}{2}$ and $m_s = \frac{1}{2}$, which in turn correspond to the two possible directions of the electron's intrinsic spin.

REVIEW CHECKLIST

▷ Understand the significance of the wave function and the associated radial probability density for the ground state of hydrogen.

▷ For each of the quantum numbers, n (the principal quantum number), ℓ (the orbital quantum number), m_ℓ (the orbital magnetic quantum number), and m_s (the spin magnetic quantum number):
  • qualitatively describe what each implies concerning atomic structure;
  • state the allowed values which may be assigned to each, and the number of allowed states that may exist in a particular atom corresponding to each quantum number.

▷ Associate the customary shell and subshell spectroscopic notations with allowed combinations of quantum numbers n and ℓ. Calculate the possible values of the orbital angular momentum, L, corresponding to a given value of the principal quantum number.

▷ Describe how allowed values of the magnetic orbital quantum number, m_ℓ, may lead to a restriction on the orientation of the orbital angular momentum vector in an external magnetic field. Find the allowed values for L_z (the component of the angular momentum along the direction of an external magnetic field) for a given value of L.

▷ State the Pauli exclusion principle and describe its relevance to the periodic table of the elements. Show how the exclusion principle leads to the known electronic ground state configuration of the light elements.

ANSWERS TO SELECTED CONCEPTUAL QUESTIONS

7. Could the Stern-Gerlach experiment be performed with ions rather than neutral atoms? Explain.

Answer

Practically speaking, the answer would be no. Since ions have a net charge, the magnetic force $q\mathbf{v} \times \mathbf{B}$ would deflect the beam, making it very difficult to separate ions with different magnetic-moment orientations.

9. Discuss some of the consequences of the exclusion principle.

Answer

If the Pauli exclusion principle were not valid, the elements and their chemical behavior would be grossly different because every electron would end up in the lowest energy level of the atom. All matter would therefore be nearly alike in its chemistry and composition, since the shell structures of each element would be identical. Most materials would have a much higher density, and the spectra of atoms and molecules would be very simple, resulting in the existence of less color in the world.

10. Why do lithium, potassium, and sodium exhibit similar chemical properties?

Answer

The three elements have similar electronic configurations, with filled inner shells plus a single electron in an s orbital. Since atoms typically interact through their unfilled outer shells, and the outer shell of each of these atoms is similar, the chemical interactions of the three atoms is also similar.

SOLUTIONS TO SELECTED END-OF-CHAPTER PROBLEMS

3. A general expression for the energy levels of one-electron atoms and ions is

$$E_n = -\frac{\mu k_e^2 q_1^2 q_2^2}{2\hbar^2 n^2}$$

where k_e is the Coulomb constant, q_1 and q_2 are the charges of the electron and the nucleus, and μ is the reduced mass, given by $\mu = m_1 m_2/(m_1 + m_2)$. In Problem 2 we found that the wavelength for the $n = 3$ to $n = 2$ transition of the hydrogen atom is 656.3 nm (visible red light). What are the wavelengths for this same transition in (a) positronium, which consists of an electron and a positron, and (b) singly ionized helium? (**Note**: A positron is a positively charged electron.)

Solution For hydrogen,
$$\mu = \frac{m_{\text{proton}} m_e}{m_{\text{proton}} + m_e} \cong m_e$$

The photon energy is
$$E_3 - E_2$$

Its wavelength is experimentally found to be
$$\lambda = \frac{c}{f} = \frac{hc}{E_3 - E_2} = 656.3 \text{ nm}$$

(a) For positronium,
$$\mu = \frac{m_e m_e}{m_e + m_e} = \frac{m_e}{2}$$

so the energy of each level is one half as large as in hydrogen, which we could call "protonium." The photon energy is inversely proportional to its wavelength, so

$$\lambda_{32} = 2(656 \text{ nm}) = 1.31 \; \mu\text{m} \quad \text{(in the infrared region)} \qquad \lozenge$$

(b) For He$^+$, $\mu \approx m_e$, $q_1 = e$, and $q_2 = 2e$, so each energy is $2^2 = 4$ times larger than hydrogen. Then,

$$\lambda_{32} = \left(\frac{656}{4}\right) \text{nm} = 164 \text{ nm} \quad \text{(in the ultraviolet region)} \qquad \lozenge$$

9. For a spherically symmetric state of a hydrogen atom, the Schrödinger equation in spherical coordinates is

$$-\frac{\hbar^2}{2m}\left(\frac{d^2\psi}{dr^2} + \frac{2}{r}\frac{d\psi}{dr}\right) - \frac{k_e e^2}{r}\psi = E\psi$$

Show that the 1s wave function for an electron in hydrogen,

$$\psi_{1s}(r) = \frac{1}{\sqrt{\pi a_0{}^3}}e^{-r/a_0}$$

satisfies the Schrödinger equation.

Solution   We use the quantum particle under boundary conditions model. We substitute the wave function and its derivatives into the Schrödinger equation, and then start simplifying the resulting equation. If we find that the resulting equation is true, then we know that the Schrödinger equation is satisfied.

$$\frac{d\psi}{dr} = \frac{1}{\sqrt{\pi a_0{}^3}}\frac{d}{dr}\left(e^{-r/a_0}\right) = -\frac{1}{\sqrt{\pi a_0{}^3}}\left(\frac{1}{a_0}\right)e^{-r/a_0} = -\frac{\psi}{a_0} \tag{1}$$

Likewise,

$$\frac{d^2\psi}{dr^2} = -\frac{1}{\sqrt{\pi a_0{}^5}}\frac{d}{dr}e^{-r/a_0} = \frac{1}{\sqrt{\pi a_0{}^7}}e^{-r/a_0} = \frac{1}{a_0{}^2}\psi \tag{2}$$

Substituting (1) and (2) into the Schrödinger equation, and noting that m is the electron's mass,

$$-\frac{\hbar^2}{2m_e}\left(\frac{1}{a_0{}^2} + \frac{2}{a_0 r}\right)\psi - \frac{k_e e^2}{r}\psi \,\psi E \tag{3}$$

Substituting $\hbar^2 = m_e k_e e^2 a_0$ (Eq. 11.21)

and canceling m_e and ψ, $-\dfrac{k_e e^2 a_0}{2}\left(\dfrac{1}{a_0{}^2} - \dfrac{2}{a_0 r}\right) - \dfrac{k_e e^2}{r} = E$

We find that the second and third terms add to zero, leaving the equation that was given for the ground state energy of hydrogen:

$$E = -\frac{k_e e^2}{2a_0}$$

This is true, so the Schrödinger equation is satisfied.   ◊

17. How many sets of quantum numbers are possible for an electron for which (a) $n = 1$, (b) $n = 2$, (c) $n = 3$, (d) $n = 4$, and (e) $n = 5$? Check your results to show that they agree with the general rule that the number of sets of quantum numbers is equal to $2n^2$.

Solution

(a)   For $n = 1$, $\ell = 0$, $m_\ell = 0$, $m_s = \pm\frac{1}{2}$.

n	ℓ	m_ℓ	m_s
1	0	0	$-\frac{1}{2}$
1	0	0	$+\frac{1}{2}$

This yields $2n^2 = 2(1)^2 = 2$ sets   ◊

(b)   For $n = 2$, we have

n	ℓ	m_ℓ	m_s
2	0	0	$\pm\frac{1}{2}$
2	1	-1	$\pm\frac{1}{2}$
2	1	0	$\pm\frac{1}{2}$
2	1	1	$\pm\frac{1}{2}$

This yields $2n^2 = 2(2)^2 = 8$ sets   ◊

Note that the number is twice the number of m_ℓ values. Also, for each ℓ there are $(2\ell + 1)$ different m_ℓ values. Finally, ℓ can take on values ranging from 0 to $n - 1$. So the general expression is

$$\text{number} = \sum_0^{n-1} 2(2\ell + 1)$$

The series is an arithmetic progression $2 + 6 + 10 + 14$,

the sum of which is $\text{number} = \dfrac{n}{2}\left[2a + (n-1)d\right]$

where $a = 2$, $d = 4$ $\text{number} = \dfrac{n}{2}\left[4 + (n-1)4\right] = 2n^2$

(c) $n = 3$: $2(1) + 2(3) + 2(5) = 2 + 6 + 10 = 18$ $2n^2 = 2(3)^2 = 18$ ◊

(d) $n = 4$: $2(1) + 2(3) + 2(5) + 2(7) = 32$ $2n^2 = 2(4)^2 = 32$ ◊

(e) $n = 5$: $32 + 2(9) = 32 + 18 = 50$ $2n^2 = 2(5)^2 = 50$ ◊

19. The ρ-meson has a charge of $-e$, a spin quantum number of 1, and a mass of 1 507 times that of the electron. Imagine that the electrons in atoms were replaced by ρ-mesons. List the possible sets of quantum numbers in the 3d subshell.

Solution   The 3d subshell has $\ell = 2$, and $n = 3$. Also, we have $s = 1$. Therefore, we can have $n = 3$, $\ell = 2$, $m_\ell = -2, -1, 0, 1, 2$, $s = 1$, and $m_s = -1, 0, 1$, leading to the following table:

n	3	3	3	3	3	3	3	3	3	3	3	3	3	3	3
ℓ	2	2	2	2	2	2	2	2	2	2	2	2	2	2	2
m_ℓ	-2	-2	-2	-1	-1	-1	0	0	0	1	1	1	2	2	2
s	1	1	1	1	1	1	1	1	1	1	1	1	1	1	1
m_s	-1	0	1	-1	0	1	-1	0	1	-1	0	1	-1	0	1

◊

27.   (a) Scanning through Table 29.4 in order of increasing atomic number, note that the electrons fill the subshells in such a way that those subshells with the lowest values of $n + \ell$ are filled first. If two subshells have the same value of $n + \ell$, the one with the lower value of n is filled first. Using these two rules, write the order in which the subshells are filled through $n + \ell = 7$. (b) Predict the chemical valence for the elements that have atomic numbers 15, 47, and 86, and compare your predictions with the actual valences.

Solution

(a)

$n + \ell$	1	2	3	4	5	6	7
subshell	1s	2s	2p, 3s	3p, 4s	3d, 4p, 5s	4d, 5p, 6s	4f, 5d, 6p, 7s

◊

(b) $Z = 15$:   Filled subshells: $1s, 2s, 2p, 3s$ (12 electrons)

　　　　　　　Valence subshell: 3 electrons in 3p subshell

　　　　　　　Prediction: Valence +3 or –5

　　　　　　　Element is phosphorus: Valence +3 or –5 (Prediction works)   ◊

　　　$Z = 47$:   Filled subshells: $1s, 2s, 2p, 3s, 3p, 4s, 3d, 4p, 5s$ (38 electrons)

　　　　　　　Outer subshell: 9 electrons in 4d subshell

　　　　　　　Prediction: Valence –1

　　　　　　　Element is silver,   Valence +1   (Prediction fails)   ◊

　　　$Z = 86$:   Filled shells: $1s, 2s, 2p, 3s, 3p, 4s, 3d, 4p, 5s, 4d, 5p, 6s, 4f, 5d, 6p$

　　　　　　　Outer subshell: Full

　　　　　　　Prediction: Inert gas

　　　　　　　Element is Radon,   Inert gas   (Prediction works)   ◊

33. Use the method illustrated in Example 29.8 to calculate the wavelength of the x-ray emitted from a molybdenum target ($Z = 42$) when an electron moves from the L shell ($n = 2$) to the K shell ($n = 1$).

Solution   Following Example 29.8, suppose the electron is originally in the L shell with just one other electron in the K shell between it and the nucleus, so it moves in a field of effective charge $(42-1)e$. Its energy is then $E_L = -(42-1)^2(13.6 \text{ eV}/4)$. In its final state we estimate the screened charge holding it in orbit as again $(42-1)e$, so its energy is $E_K = -(42-1)^2(13.6 \text{ eV})$. The photon energy emitted is the difference.

$$E_\gamma = \frac{3}{4}(42-1)^2(13.6 \text{ eV}) = 1.71\times 10^4 \text{ eV} = 2.74\times 10^{-15} \text{ J}$$

Then $f = \dfrac{E}{h} = 4.14\times 10^{18}$ Hz   and $\lambda = \dfrac{c}{f} = 72.5$ pm   ◊

---

39. A distant quasar is moving away from Earth at such high speed that the blue 434-nm H_γ line of hydrogen is observed at 510 nm, in the green portion of the spectrum (Fig. P29.39). (a) How fast is the quasar receding? You may use the result of Problem 38. (b) Edwin Hubble discovered that all objects outside the local group of galaxies are moving away from us, with speeds proportional to their distances. Hubble's law is expressed as $v = HR$, where Hubble's constant has the approximate value 17×10^{-3} m/s·ly. Determine the distance from Earth to this quasar.

Solution   The problem states that the quasar is moving very fast, and since there is a significant red shift of the light, the quasar must be moving away from Earth at a relativistic speed ($v > 0.1c$). Quasars are very distant astronomical objects, and since our universe is estimated to be about 15 billion years old, we should expect this quasar to be $\sim 10^9$ light-years away.

As suggested, we can use the equation in Problem 38 to find the speed of the quasar from the Doppler red shift, and this speed can then be used to find the distance using Hubble's law.

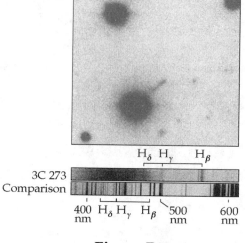

Figure P29.39

(a) $\dfrac{\lambda'}{\lambda} = \dfrac{510\text{ nm}}{434\text{ nm}} = 1.18 = \sqrt{\dfrac{1+v/c}{1-v/c}}$

Squared, this becomes $\qquad \dfrac{1+v/c}{1-v/c} = 1.38 \quad$ or $\quad 2.38v/c = 0.381$

Therefore, $\qquad\qquad\qquad\qquad v = 0.160c \qquad$ or $\qquad 16.0\%$ of the speed of light $\Diamond$

(b) Hubble's law asserts that the universe is expanding at a constant rate so that the speeds of galaxies are proportional to their distance R from Earth, $v = HR$.

So, $\qquad\qquad\qquad\qquad R = \dfrac{v}{H} = \dfrac{(0.160)\left(3.00 \times 10^8 \text{ m/s}\right)}{\left(1.70 \times 10^{-2} \text{ m/s} \cdot \text{ly}\right)} = 2.82 \times 10^9 \text{ ly} \qquad \Diamond$

---

41. Show that the average value of r for the 1s state of hydrogen has the value $\dfrac{3}{2}a_0$. (**Hint:** Use Eq. 29.7.)

Solution

The average value (expectation value) of r is $\qquad\qquad <r> = \int_0^\infty r\, P_{1s}(r)\, dr$

where $P_{1s}(r) = \left(\dfrac{4r^2}{a_0^3}\right) e^{-2r/a_0}$: $\qquad\qquad <r> = \dfrac{4}{a_0^3} \int_0^\infty r^3 e^{-2r/a_0}\, dr$

Letting $x = \dfrac{2r}{a_0}$ we find: $\qquad\qquad\qquad <r> = \dfrac{1}{4} a_0 \int_0^\infty x^3 e^{-x}\, dx$

Integrating by parts gives $\qquad\qquad\qquad <r> = \dfrac{3}{2} a_0 \qquad\qquad \Diamond$

---

43. Suppose a hydrogen atom is in the 2s state. Taking $r = a_0$, calculate values for (a) $\psi_{2s}(a_0)$, (b) $|\psi_{2s}(a_0)|^2$, and (c) $P_{2s}(a_0)$. (**Hint:** Use Eq. 29.8.)

Solution   The wave function for the 2s state is given by Equation 29.8:

$$\psi_{2s}(r) = \frac{1}{4\sqrt{2\pi}}\left(\frac{1}{a_0}\right)^{3/2}\left(2 - \frac{r}{a_0}\right)e^{-r/2a_0}$$

(a) Taking $r = a_0 = 0.529 \times 10^{-10}$ m, we find

$$\psi_{2s}(a_0) = \frac{1}{4\sqrt{2\pi}}\left(\frac{1}{0.529 \times 10^{-10} \text{ m}}\right)^{3/2}(2-1)e^{-1/2} = 1.57 \times 10^{14} \text{ m}^{-3/2} \qquad \Diamond$$

(b) $|\psi_{2s}(a_0)| = \left(1.57 \times 10^{14} \text{ m}^{-3/2}\right)^2 = 2.47 \times 10^{28} \text{ m}^{-3} \qquad \Diamond$

(c) Using Equation 29.5 and the results to (b) gives

$$P_{2s}(a_0) = 4\pi a_0^2 |\psi_{2s}(a_0)|^2 = 8.69 \times 10^8 \text{ m}^{-1} \qquad \Diamond$$

---

45. (a) Show that the most probable radial position for an electron in the 2s state of hydrogen is $r = 5.326a_0$. (b) Show that the wave function given by Equation 29.8 is normalized.

Solution

We are using the quantum particle model.

We use Equation 29.8: $\psi_{2s}(r) = \frac{1}{4\sqrt{2\pi}}\left(\frac{1}{a_0}\right)^{3/2}\left[2 - \frac{r}{a_0}\right]e^{-r/2a_0}$

By Equation 29.6, the radial probability distribution function is

$$P(r) = 4\pi r^2 \psi^2 = \frac{1}{8}\left(\frac{r^2}{a_0^3}\right)\left(2 - \frac{r}{a_0}\right)^2 e^{-r/a_0}$$

(a) Its extremes are given by

$$\frac{dP(r)}{dr} = \frac{1}{8}\left[\frac{2r}{a_0^3}\left(2-\frac{r}{a_0}\right)^2 - \frac{2r^2}{a_0^3}\left(\frac{1}{a_0}\right)\left(2-\frac{r}{a_0}\right) - \frac{r^2}{a_0^3}\left(2-\frac{r}{a_0}\right)^2\left(\frac{1}{a_0}\right)\right]e^{-r/a_0} = 0$$

We factor in the following manner:

$$\frac{1}{8}\left(\frac{r}{a_0^3}\right)\left(2-\frac{r}{a_0}\right)\left[2\left(2-\frac{r}{a_0}\right) - \frac{2r}{a_0} - \frac{r}{a_0}\left(2-\frac{r}{a_0}\right)\right]e^{-r/a_0} = 0$$

The roots of $\frac{dP}{dr} = 0$ at $r = 0$, $r = 2a_0$, and $r = \infty$ are minima $(P(r) = 0)$.

Therefore, we focus on the roots given by

$$2\left(2-\frac{r}{a_0}\right) - 2\frac{r}{a_0} - \left(\frac{r}{a_0}\right)\left(2-\frac{r}{a_0}\right) = 4 - \frac{6r}{a_0} + \left(\frac{r}{a_0}\right)^2 = 0$$

which has solutions $r = \left(3 \pm \sqrt{5}\right)a_0$

We substitute two roots into $P(r)$:

When $r = \left(3 - \sqrt{5}\right)a_0 = 0.764\,a_0$, then $P(r) = \dfrac{0.0519}{a_0}$

When $r = \left(3 + \sqrt{5}\right)a_0 = 5.24\,a_0$, then $P(r) = \dfrac{0.191}{a_0}$

Therefore, the most probable value of r is $\left(3 + \sqrt{5}\right)a_0 = 5.24\,a_0$ ◊

(b) From Equation 29.8, $\displaystyle\int_0^\infty P(r)\,dr = \int_0^\infty \frac{1}{8}\left(\frac{r^2}{a_0^3}\right)\left(2-\frac{r}{a_0}\right)^2 e^{-r/a_0}\,dr$

Let $u = \dfrac{r}{a_0}$, and $dr = a_0\,du$, so that $\displaystyle\int_0^\infty P(r)\,dr = \int_0^\infty \frac{1}{8}u^2(4 - 4u + u^2)e^{-u}\,du$

$$\int_0^\infty P(r)\,dr = \int_0^\infty \frac{1}{8}\left(u^4 - 4u^3 + 4u^2\right)e^{-u}\,du$$

Using a table of integrals, or integrating by parts repeatedly,

$$\int_0^\infty P(r)\,dr = -\frac{1}{8}\left(u^4 + 4u^2 + 8u + 8\right)e^{-u}\Big|_0^\infty = 1 \text{ as desired}$$ ◊

47. An electron in chromium moves from the $n = 2$ state to the $n = 1$ state without emitting a photon. Instead, the excess energy is transferred to an outer electron (one in the $n = 4$ state), which is then ejected by the atom. (This is called an Auger [pronounced 'ohjay'] process, and the ejected electron is referred to as an Auger electron.) Use the Bohr theory to find the kinetic energy of the Auger electron.

Solution

The chromium atom with nuclear charge $Z = 24$ starts with one vacancy in the $n = 1$ shell, perhaps produced by the absorption of an x-ray, which ionized the atom. An electron from the $n = 2$ shell tumbles down to fill the vacancy. We suppose that the electron is shielded from the electric field of the full nuclear charge by the one K-shell electron originally below it. Its change in energy is

$$\Delta E = -(Z-1)^2(13.6 \text{ eV})\left(\frac{1}{1^2} - \frac{1}{2^2}\right) = -5.40 \text{ keV}$$

Then $+5.40$ keV can be transferred to the single $4s$ electron. Suppose that it is shielded by the 22 electrons in the K, L, and M shells. To break the outermost electron out of the atom, producing a Cr^{2+} ion, requires an energy investment of

$$E_{\text{ionize}} = \frac{(Z-22)^2(13.6 \text{ eV})}{4^2} = \frac{2^2(13.6 \text{ eV})}{16} = 3.40 \text{ eV}$$

As evidence that this relatively tiny amount of energy can still be the right order of magnitude, note that the (first) ionization energy for neutral chromium is tabulated as 6.76 eV. Then the remaining energy that can appear as kinetic energy is

$$K = \Delta E - E_{\text{ionize}} = 5395.8 \text{ eV} - 3.4 \text{ eV} = 5.39 \text{ keV} \qquad \lozenge$$

Because of conservation of momentum for the ion-electron system and the tiny mass of the electron compared to that of the Cr^{2+} ion, almost all of this kinetic energy will belong to the electron.

51. According to classical physics, a charge e moving with an acceleration a radiates at a rate

$$\frac{dE}{dt} = -\frac{1}{6\pi\epsilon_0}\frac{e^2 a^2}{c^3}$$

(a) Show that an electron in a classical hydrogen atom (see Fig. 29.3) spirals into the nucleus at a rate

$$\frac{dr}{dt} = -\frac{e^4}{12\pi^2\epsilon_0{}^2 r^2 m_e{}^2 c^3}$$

(b) Find the time it takes the electron to reach $r = 0$, starting from $r_0 = 2.00 \times 10^{-10}$ m.

Solution   According to a classical model, the electron moving as a particle in uniform circular motion about the proton in the hydrogen atom experiences a force $k_e e^2 / r^2$; and from Newton's second law, $F = ma$, its acceleration is $k_e e^2 / m_e r^2$.

(a) Using the fact that the Coulomb constant $k_e = \dfrac{1}{4\pi\epsilon_0}$

$$a = \frac{v^2}{r} = \frac{k_e e^2}{m_e r^2} = \frac{e^2}{4\pi\epsilon_0 m_e r^2} \qquad (1)$$

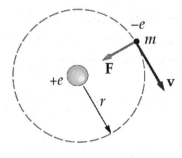

Figure 29.3

From the Bohr structural model of the atom (Chapter 11), we can write the total energy of the atom as

$$E = -\frac{k_e e^2}{2r} = -\frac{e^2}{8\pi\epsilon_0 r}$$

so

$$\frac{dE}{dt} = \frac{e^2}{8\pi\epsilon_0 r^2}\frac{dr}{dt} = -\frac{1}{6\pi\epsilon_0}\frac{e^2 a^2}{c^3} \qquad (2)$$

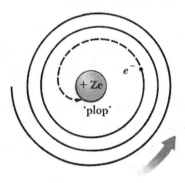

Substituting (1) into (2) for a, solving for $\dfrac{dr}{dt}$, and simplifying gives

$$\frac{dr}{dt} = -\frac{4r^2}{3c^3}\left(\frac{e^2}{4\pi\epsilon_0 m_e r^2}\right)^2 = -\frac{e^4}{12\pi^2\epsilon_0{}^2 r^2 m_e{}^2 c^3} \qquad \lozenge$$

(b) We can express $\dfrac{dr}{dt}$ in the simpler form:

$$\frac{dr}{dt} = -\frac{A}{r^2} = -\frac{3.15 \times 10^{-21}}{r^2}$$

Thus,

$$-\int_{2.00 \times 10^{-10} \text{ m}}^{0} r^2 \, dr = 3.15 \times 10^{-21} \int_{0}^{T} dt$$

and

$$T = \left(3.17 \times 10^{20}\right) \frac{r^3}{3} \Bigg|_{0}^{2.00 \times 10^{-10} \text{ m}}$$

$$T = 8.46 \times 10^{-10} \text{ s} = 0.846 \text{ ns} \qquad \Diamond$$

We know that atoms last much longer than 0.8 ns; thus, classical physics does not hold (fortunately) for atomic systems.

Chapter 30

Nuclear Physics

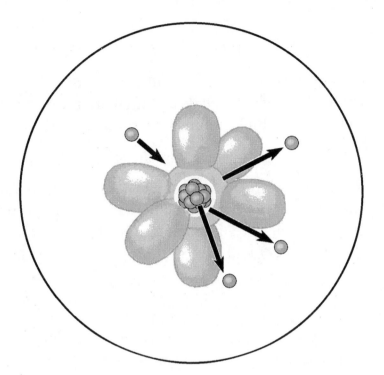

INTRODUCTION

In this chapter we discuss the properties and structure of the atomic nucleus. We start by describing the basic properties of nuclei, and then discuss nuclear forces and binding energy, nuclear models, and the phenomenon of radioactivity. We also discuss nuclear reactions and the various processes by which nuclei decay.

NOTES FROM SELECTED CHAPTER SECTIONS

30.1   Some Properties of Nuclei

Important quantities in the description of nuclear properties are:

- The **atomic number**, Z, which equals the number of protons in the nucleus.
- The **neutron number**, N, which equals the number of neutrons in the nucleus.
- The **mass number**, A, which equals the number of nucleons (neutrons plus protons) in the nucleus.

The nuclei of all atoms of a particular element contain the same number of protons but often contain different numbers of neutrons. Nuclei that are related in this way are called **isotopes**. The isotopes of an element have the same Z value but different N and A values.

The **atomic mass unit**, u, is defined such that the mass of one atom of the isotope ^{12}C is exactly 12 u.

Experiments have shown that most nuclei can be geometrically modeled as being approximately spherical and all **have nearly the same density**. The stability of nuclei is due to the **nuclear force.** This is a **short range**, **attractive** force, which acts between all nuclear particles. Nuclei have **intrinsic angular momentum** which is quantized by the **nuclear spin quantum number**, which may be integer or half integer.

The **magnetic moment of the nucleus** is measured in terms of the **nuclear magneton**. When placed in an external magnetic field, nuclear magnetic moments precess with a frequency called the **Larmor precessional frequency.**

30.2   Binding Energy

The total rest energy of the nucleus is always less than the combined rest energies of its individual nucleons. The binding energy of the nucleus is the difference in these two energy values.

30.3   Radioactivity

Radioactive substances can emit three possible types of radiation by spontaneous decay:

- alpha particles(α); nuclei of Helium atoms
- beta particles (β); either electrons or positrons
- gamma rays (γ); high-energy photons

A positron is a particle similar to the electron in all respects except that it has a charge of $+e$, and is the antimatter twin of the electron. In beta decay, the electron is denoted by a e^-, while the positron is designated with a e^+.

The three types of radiation have quite different penetrating powers. Alpha particles barely penetrate a sheet of paper, beta particles can penetrate a few millimeters of aluminum, and gamma rays can penetrate several centimeters of lead.

The half-life of a radioactive substance is the time required for half of a given number of radioactive nuclei to decay.

30.4   The Radioactive Decay Processes

Alpha decay can occur because, according to quantum mechanics, some nuclei have energy barriers that can be penetrated by the alpha particles (the tunneling process). Beta decay is energetically more favorable for those nuclei having a large excess of neutrons. A nucleus can undergo beta decay in two ways. It can emit either an electron (e^-) and an antineutrino ($\bar{v}$), or a positron (e^+) and a neutrino (v). In the electron-capture process, the nucleus of an atom absorbs one of its own electrons (usually from the K shell) and emits a neutrino.

The neutrino has the following properties:

- It has zero electric charge.
- It has a mass smaller than that of the electron, and in fact its mass may be zero (although recent experiments suggest that this may not be true).
- It has a spin of 1/2, which satisfies the law of conservation of angular momentum.
- It interacts very weakly with matter and is therefore very difficult to detect.

In gamma decay, a nucleus in an excited state decays to its ground state and emits a gamma ray.

The disintegration energy is the energy released as a result of the decay process.

30.5   Nuclear Reactions

Nuclear reactions are events in which collisions with energetic particles change the identity or properties of nuclei. The total change in the rest energy as a result of a nuclear reaction is called the **reaction energy**, Q.

An **endothermic reaction** is one in which Q is negative, and the minimum energy for which the reaction will occur is called the **threshold energy.**

EQUATIONS AND CONCEPTS

Most nuclei are approximately spherical in shape and have an average radius that is proportional to the cube root of the mass number, or total number of nucleons. This means that the volume is proportional to A and that all nuclei have nearly the same density.

$$r = r_0 A^{1/3} \tag{30.1}$$

$$r_0 = 1.2 \times 10^{-15} \text{ m} = 1.2 \text{ fm}$$

The nucleus has an angular momentum and a corresponding nuclear magnetic moment associated with it. The nuclear magnetic moment is measured in terms of a unit of moment called the nuclear magneton μ_n.

$$\mu_n \equiv \frac{e\hbar}{2m_p} = 5.05 \times 10^{-27} \text{ J/T} \tag{30.3}$$

The **binding energy** of any nucleus can be calculated in terms of the mass of a neutral hydrogen atom, the mass of a neutron, and the atomic mass of the associated compound nucleus.

$$E_b(\text{MeV}) = \left(ZM(\text{H}) + Nm_n - M({}^A_Z\text{X}) \right)\left(931.494 \, \frac{\text{MeV}}{\text{u}} \right) \tag{30.4}$$

The number of radioactive nuclei in a given sample which undergoes decay during a time interval Δt depends on the number of nuclei present. The number of decays depends also on the **decay constant** λ, which is characteristic of a particular isotope.

$$\frac{dN}{dt} = -\lambda N \tag{30.5}$$

The number of nuclei in a radioactive sample decreases exponentially with time. The plot of number of nuclei N versus elapsed time t is called a decay curve.

$$N = N_0 e^{-\lambda t} \qquad (30.6)$$

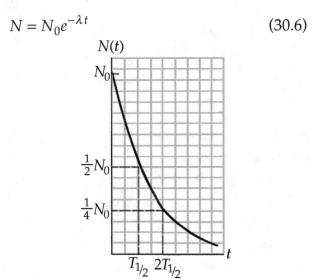

The **decay rate** R or activity of a sample of radioactive nuclei is defined as the number of decays per second.

$$R = \left| \frac{dN}{dt} \right| = N_0 \lambda e^{-\lambda t} = R_0 e^{-\lambda t} \qquad (30.7)$$

The **half-life** $T_{1/2}$ is the time required for half of a given number of radioactive nuclei to decay.

$$T_{1/2} = \frac{\ln 2}{\lambda} = \frac{0.693}{\lambda} \qquad (30.8)$$

When a nucleus decays by **alpha emission**, the parent nucleus loses two neutrons and two protons. In order for alpha emission to occur, the mass of the parent nucleus must be greater than the combined mass of the daughter nucleus and the emitted alpha particle. The mass difference is converted into energy and appears as kinetic energy shared (unequally) by the alpha particle and the daughter nucleus.

$$^{A}_{Z}X \rightarrow ^{A-4}_{Z-2}Y + ^{4}_{2}He \qquad (30.9)$$

$$^{238}_{92}U \rightarrow ^{234}_{90}Th + ^{4}_{2}He \qquad (30.10)$$

The **disintegration energy** Q can be calculated in MeV when the masses are expressed in u.

$$Q = (M_X - M_Y - M_\alpha)(931.494 \text{ MeV}/\text{u}) \qquad (30.13)$$

When a radioactive nucleus undergoes **beta decay**, the daughter nucleus has the same mass number as the parent nucleus, but the charge number (or atomic number) increases by one. The electron that is emitted is created within the parent nucleus by a process, which can be modeled by a neutron transformed into a proton and an electron. The total energy released in beta decay is greater than the combined kinetic energies of the electron and the daughter nucleus. This difference in energy is associated with a third particle called a neutrino.

$$_Z^A X \rightarrow _{Z+1}^A Y + e^- + \bar{v} \qquad (30.16)$$

$$n \rightarrow p + e^- + \bar{v}$$

Electron capture occurs when a parent nucleus captures one of its own orbital electrons and emits a neutrino. This process is characteristic of neutron-rich nuclei.

$$_Z^A X + _{-1}^0 e \rightarrow _{Z-1}^A Y + v \qquad (30.19)$$

Nuclei which undergo alpha or beta decay are often left in an excited energy state. The nucleus returns to the ground state by emission of one or more photons. Gamma-ray emission results in no change in mass number or atomic number.

$$_Z^A X^* \rightarrow _Z^A X + \gamma \qquad (30.20)$$

Nuclear reactions can occur when target nuclei are bombarded with energetic particles. In these reactions the structure, identity, or properties of the target nuclei are changed.

$$a + X \rightarrow Y + b \qquad (30.23)$$

The **reaction energy** Q associated with a nuclear reaction is the total change in rest energy which results from the reaction. Q is positive for an **exothermic reaction** and negative for an **endothermic reaction**.

$$Q = (M_a + M_X - M_Y - M_b)c^2 \qquad (30.24)$$

TABLE 30.2   Various Decay Pathways

Alpha Decay	$^A_Z X \rightarrow ^{A-4}_{Z-2} Y + ^4_2 He$
Beta Decay (e^-)	$^A_Z X \rightarrow ^A_{Z+1} Y + e^- + \bar{\nu}$
Beta Decay (e^+)	$^A_Z X \rightarrow ^A_{Z-1} Y + e^+ + \nu$
Electron Capture	$^A_Z X + ^{\,0}_{-1} e \rightarrow ^A_{Z-1} Y + \nu$
Gamma Decay	$^A_Z X^* \rightarrow ^A_Z X + \gamma$

SUGGESTIONS, SKILLS, AND STRATEGIES

The rest energy of a particle is given by $E = mc^2$. It is therefore often convenient to express the unified mass unit in terms of its equivalent energy, $1\,u = 1.660559 \times 10^{-27}$ kg or $1\,u = 931.494\ \text{MeV}/c^2$. When masses are expressed in units of u, energy values are then $E = m(931.494\ \text{MeV}/u)$.

Equation 30.6 can be solved for the particular time t after which the number of remaining nuclei will be some specified fraction of the original number N_0. This can be done by taking the natural log of each side of Equation 30.6 to find

$$t = \frac{1}{\lambda} \ln\left(\frac{N_0}{N}\right)$$

REVIEW CHECKLIST

▷ Use the appropriate nomenclature in describing the static properties of nuclei.

▷ Discuss nuclear stability in terms of the strong nuclear force and a plot of N vs. Z.

▷ Account for nuclear binding energy in terms of the Einstein mass-energy relationship. Describe the basis for energy released by fission and fusion in terms of the shape of the curve of binding energy per nucleon vs. mass number.

▷ Identify each of the components of radiation that are emitted by the nucleus through natural radioactive decay and describe the basic properties of each. Write out generic equations (e.g. Eqs. 30.9, 30.16, 30.20, and 30.19) that illustrate the conservation of nucleon number and charge in the processes of radioactive decay by alpha, beta and gamma emission and by electron capture. Explain why the neutrino must be considered in the analysis of beta decay.

▷ State and apply to the solution of related problems, the formula which expresses decay rate as a function of the decay constant and the number of radioactive nuclei. Describe the process of carbon dating as a means of determining the age of ancient objects.

▷ Calculate the Q value of given nuclear reactions and determine the threshold energy of endothermic reactions.

ANSWERS TO SELECTED CONCEPTUAL QUESTIONS

5.   Why do nearly all the naturally occurring isotopes lie above the $N = Z$ line in Figure 30.4?

Answer   As Z increases, extra neutrons are required to overcome the increasing electrostatic repulsion of the protons.

□   □   □   □

8.   Two samples of the same radioactive nuclide are prepared. Sample A has twice the initial activity of sample B. How does the half-life of A compare with the half-life of B? After each has passed through five half-lives, what is the ratio of their activities?

Answer   Since the two samples are of the same radioactive nuclide, they have the same half-life; the 2 : 1 difference in activity is due to a 2 : 1 difference in the mass of each sample. After 5 half lives, each will have decreased in mass by a power of $2^5 = 32$. However, since this simply means that the mass of each is 32 times smaller, the ratio of the masses will still be (2/32) : (1/32), or 2 : 1. Therefore, the ratio of their activities will **always** be 2 : 1.

□   □   □   □

12.   If a nucleus such as ^{226}Ra initially at rest undergoes alpha decay, which has more kinetic energy after the decay, the alpha particle or the daughter nucleus?

Answer   The alpha particle and the daughter nucleus carry momenta of equal magnitudes in opposite directions. Since kinetic energy can be written as $p^2/2m$, the less massive alpha particle has much more of the decay energy than the recoiling nucleus.

□   □   □   □

16.   Suppose it could be shown that the cosmic ray intensity at the Earth's surface was much greater 10 000 years ago. How would this difference affect what we accept as valid carbon-dated values of the age of ancient samples of once-living matter?

Answer   If the cosmic ray intensity at the Earth's surface was much greater 10 000 years ago, the fraction of the Earth's carbon dioxide with the heavy nuclide ^{14}C would also be greater at that time, than now. Thus, there would initially be a greater fraction of ^{14}C in the organic artifacts, and we would believe that the artifact was more recent than it actually was.

For example, suppose that the actual ratio of atmospheric ^{14}C to ^{12}C, two half-lives (11 460 years) ago was 2.6×10^{-12}. The current ratio of isotopes would be

$$\left(\tfrac{1}{2}\right)\left(\tfrac{1}{2}\right)\left(2.6 \times 10^{-12}\right) = 0.65 \times 10^{-12}$$

We would still measure this ratio today; but, believing the initial ratio to be 1.3×10^{-12}, we would think that the creature had died only one half-life (5730 years) ago.

$$\tfrac{1}{2}\left(1.3 \times 10^{-12}\right) = 0.65 \times 10^{-12}$$

□   □   □   □

SOLUTIONS TO SELECTED END-OF-CHAPTER PROBLEMS

5. A star ending its life with a mass of two times the mass of the Sun is expected to collapse, combining its protons and electrons to form a neutron star. Such a star could be thought of as a gigantic atomic nucleus. If a star of mass $2 \times 1.99 \times 10^{30}$ kg collapsed into neutrons ($m_n = 1.67 \times 10^{-27}$ kg), what would the radius be? (Assume $r = r_0 A^{1/3}$.)

Solution The number of nucleons in the star is

$$A = \frac{2 \times 1.99 \times 10^{30} \text{ kg}}{1.67 \times 10^{-27} \text{ kg}} = 2.38 \times 10^{57}$$

Therefore, $\quad r = r_0 A^{1/3} = \left(1.20 \times 10^{-15} \text{ m}\right)\left(2.38 \times 10^{57}\right)^{1/3} = 16.0 \text{ km} \quad\quad\Diamond$

9. Using the graph in Figure 30.10, estimate how much energy is released when a nucleus of mass number 200 fissions into two nuclei each of mass number 100.

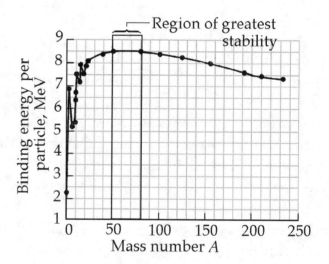

Solution The curve of binding energy shows that a heavy nucleus of mass number $A = 200$ has total binding energy about

(7.4 MeV/nucleon)(200 nucleons) = 1.5 GeV

Thus, it is less stable than its potential fission products, two middleweight nuclei of $A = 100$, having total binding energy

$2(8.4 \text{ MeV/nucleon})(100 \text{ nucleons}) = 1.7 \text{ GeV}$

Figure 30.10

Fission then releases about $\quad\quad$ 1.7 GeV – 1.5 GeV = 200 MeV $\quad\quad\quad\Diamond$

This is the energy source of uranium bombs and of nuclear electric-generating plants.

11. A pair of nuclei for which $Z_1 = N_2$ and $Z_2 = N_1$ are called **mirror isobars** (the atomic and neutron numbers are interchanged). Binding energy measurements on these nuclei can be used to obtain evidence of the charge independence of nuclear forces (i.e., proton-proton, proton-neutron, and neutron-neutron nuclear forces are equal). Calculate the difference in binding energy for the two mirror isobars $^{15}_{8}O$ and $^{15}_{7}N$. The electric repulsion among eight protons rather than seven accounts for the difference.

Solution

For $^{15}_{8}O$ we have   (using Equation 30.4)

$$E_b = \left[8(1.007\,825) + 7(1.008\,665)\,u - 15.003\,065\,u\right](931.5\ \text{MeV}/u) = 111.96\ \text{MeV}$$

For $^{15}_{7}N$ we have

$$E_b = \left[7(1.007\,825) + 8(1.008\,665)\,u - 15.000\,108\,u\right](931.5\ \text{MeV}/u) = 115.49\ \text{MeV}$$

Therefore, the difference in the two binding energies is $\Delta E_b = 3.54\ \text{MeV}$ ◊

---

15. A freshly prepared sample of a certain radioactive isotope has an activity of 10.0 mCi. After 4.00 h, its activity is 8.00 mCi. (a) Find the decay constant and half-life. (b) How many atoms of the isotope were contained in the freshly prepared sample? (c) What is the sample's activity 30.0 h after it is prepared?

Solution   Over the course of 4 hours, this isotope lost 20% of its activity, so its half-life appears to be around 10 hours, which means that its activity after 30 hours (~3 half-lives) will be about 1 mCi. The decay constant and number of atoms are not so easy to estimate.

From the rate equation, $R = R_0 e^{-\lambda t}$, we can find the decay constant λ, which can then be used to find the half-life, the original number of atoms, and the activity at any other time, t.

(a) $\lambda = \dfrac{1}{t}\ln\left(\dfrac{R_0}{R}\right) = \left(\dfrac{1}{(4.00\ \text{h})(3600\ \text{s}/\text{h})}\right)\ln\left(\dfrac{10.0\ \text{mCi}}{8.00\ \text{mCi}}\right) = 1.55\times10^{-5}\ \text{s}^{-1}$ ◊

$$T_{1/2} = \dfrac{\ln 2}{\lambda} = \dfrac{0.693}{0.0558\ \text{h}^{-1}} = 12.4\ \text{h}$$ ◊

(b) The number of original atoms can be found if we convert the initial activity from curies into becquerels (decays per second): $1 \text{ Ci} = 3.7 \times 10^{10}$ Bq.

$$R_0 = 10.0 \text{ mCi} = \left(10.0 \times 10^{-3} \text{ Ci}\right)\left(3.70 \times 10^{10} \text{ Bq/Ci}\right) = 3.70 \times 10^8 \text{ Bq}$$

Since $R_0 = \lambda N_0$, $\quad N_0 = \dfrac{R_0}{\lambda} = \dfrac{3.70 \times 10^8 \text{ decays/s}}{1.55 \times 10^{-5} \text{ s}^{-1}} = 2.39 \times 10^{13}$ atoms ◊

(c) $\quad R = R_0 e^{-\lambda t} = (10.0 \text{ mCi}) e^{-(5.58 \times 10^{-2} \text{ h}^{-1})(30.0 \text{ h})} = 1.88 \text{ mCi}$ ◊

---

19. A building has become accidentally contaminated with radioactivity. The longest-lived material in the building is strontium-90 ($^{90}_{38}$Sr has an atomic mass 89.907 7 u, and its half-life is 29.1 years. It is particularly dangerous because it substitutes for calcium in bones.) Assume that the building initially contained 5.00 kg of this substance uniformly distributed throughout the building (a very unlikely situation) and the safe level is defined as less than 10.0 decays/min (to be small in comparison with the background radiation). How long will the building be unsafe?

Solution The number of nuclei in the original sample is

$$N_0 = \frac{\text{mass present}}{\text{mass of nucleus}} = \frac{5.00 \text{ kg}}{(89.9077 \text{ u})\left(1.66 \times 10^{-27} \text{ kg/u}\right)} = 3.35 \times 10^{25} \text{ nuclei}$$

$$\lambda = \frac{\ln 2}{T_{1/2}} = \frac{0.693}{29.1 \text{ yr}} = 2.38 \times 10^{-2} \text{ yr}^{-1} = 4.52 \times 10^{-8} \text{ min}^{-1}$$

$$R_0 = \lambda N_0 = \left(4.52 \times 10^{-8} \text{ min}^{-1}\right)\left(3.35 \times 10^{25} \text{ nuclei}\right) = 1.52 \times 10^{18} \text{ counts/min}$$

$$\frac{R}{R_0} = \frac{10.0 \text{ counts/min}}{1.52 \times 10^{18} \text{ counts/min}} = 6.599 \times 10^{-18} = e^{-\lambda t}$$

$$t = \frac{-\ln(R/R_0)}{\lambda} = \frac{-\ln\left(6.599 \times 10^{-18}\right)}{2.38 \times 10^{-2} \text{ yr}^{-1}} = 1660 \text{ yr}$$ ◊

---

21. Find the energy released in the alpha decay $^{238}_{92}\text{U} \rightarrow ^{234}_{90}\text{Th} + ^{4}_{2}\text{He}$. You will find Table 30.3 useful.

Solution Table 30.3 contains the following values:

$$M(^{238}_{92}\text{U}) = 238.050\,784 \text{ u} \qquad M(^{234}_{90}\text{Th}) = 234.043\,593 \text{ u} \qquad M(^{4}_{2}\text{He}) = 4.002\,602 \text{ u}$$

We apply to the decay a relativistic energy version of the isolated system model. Some of the rest energy of the parent nucleus is converted into kinetic energy of the decay products according to $Q = \Delta mc^2$.

We calculate

$$Q = (M_\text{U} - M_\text{Th} - M_\text{He})(931.5 \text{ MeV/u})$$

$$Q = (238.050\,784 - 234.043\,593 - 4.002\,602)(931.5) = 4.27 \text{ MeV} \qquad \Diamond$$

---

25. The nucleus $^{15}_{8}\text{O}$ decays by electron capture. The nuclear reaction is written $^{15}_{8}\text{O} + e^- \rightarrow ^{15}_{7}\text{N} + \nu$. (a) Write the process going on for a single particle within the nucleus. (b) Write the decay process referring to neutral atoms. (c) Determine the energy of the neutrino. Disregard the daughter's recoil.

Solution

(a) $e^- + p \rightarrow n + \nu$ $\qquad\qquad\qquad\qquad\qquad\qquad\qquad\qquad\qquad\qquad\qquad \Diamond$

(b) Add 7 protons, 7 neutrons, and 7 electrons to each side to give

$$^{15}\text{O atom} \rightarrow ^{15}\text{N atom} + \nu \qquad\qquad\qquad\qquad\qquad \Diamond$$

(c) From Table A.3, $M(^{15}\text{O}) = M(^{15}\text{N}) + \dfrac{Q}{c^2}$

$$\Delta M = 15.003\,065 - 15.000\,108 = 0.002\,957 \text{ u}$$

$$Q = (931.5 \text{ MeV/u})(0.002\,957 \text{ u}) = 2.75 \text{ MeV} \qquad\qquad\qquad \Diamond$$

---

29. Natural gold has only one isotope, $^{197}_{79}\text{Au}$. If natural gold is irradiated by a flux of slow neutrons, electrons are emitted. (a) Write the reaction equation. (b) Calculate the maximum energy of the emitted electrons. The mass of $^{198}_{80}\text{Hg}$ is 197.966743 u.

Solution

The $^{197}_{79}\text{Au}$ will absorb a neutron to become $^{198}_{79}\text{Au}$, which emits an e^- to become $^{198}_{80}\text{Hg}$.

(a) For nuclei, the reaction is:

$^{197}_{79}\text{Au nucleus} + ^1_0\text{n} \rightarrow ^{198}_{80}\text{Hg nucleus} + ^0_{-1}\text{e}^- + \overline{\nu}$ ◊

(b) Adding 79 electrons to both sides:

$^{197}_{79}\text{Au atom} + ^1_0\text{n} \rightarrow ^{198}_{80}\text{Hg atom} + \overline{\nu}$

From Table A.3,

$$196.966\,543 + 1.008\,665 = 197.966\,743 + 0 + \frac{Q}{c^2}$$

$$Q = mc^2 = (0.008\,465\ \text{u})(931.5\ \text{MeV/u})$$

$$Q = 7.89\ \text{MeV}$$ ◊

---

33. Review Problem. Suppose enriched uranium containing 3.40% of the fissionable isotope $^{235}_{92}\text{U}$ is used to fuel a ship. The water exerts an average frictional force of magnitude 1.00×10^5 N on the ship. How far can the ship travel per kilogram of fuel? Assume that the energy released per fission event is 208 MeV and that the ship's engine has an efficiency of 20.0%.

Solution   Nuclear fission is much more efficient for converting mass to energy than burning fossil fuels. However, without knowing the rate of diesel fuel consumption for a comparable ship, it is difficult to estimate the nuclear fuel rate. It seems plausible that a ship could cross the Atlantic ocean with only a few kilograms of nuclear fuel, so a reasonable range of uranium fuel consumption might be 10 km/kg to 10 000 km/kg.

The fuel consumption rate can be found from the energy released by the nuclear fuel and the work required to push the ship through the water. We use the particle in equilibrium model. The thrust force P exerted on the propeller by the water around it must be equal in magnitude to the backward frictional (drag) force on the hull.

One kg of enriched uranium contains 3.40% $^{235}_{92}U$

so
$$m_{235} = (1000 \text{ g})(0.0340) = 34.0 \text{ g}$$

In terms of number of nuclei,

this is equivalent to
$$N_{235} = \left(\frac{34.0 \text{ g}}{235 \text{ g/mol}}\right)(6.02 \times 10^{23} \text{ atoms/mol}) = 8.71 \times 10^{22} \text{ nuclei}$$

If all these nuclei fission, the internal energy released is equal to

$$(8.71 \times 10^{22} \text{ nuclei})(208 \text{ MeV/nucleus})(1.602 \times 10^{-19} \text{ J/eV}) = 2.90 \times 10^{12} \text{ J}$$

Now, for the engine,

$$\text{efficiency} = \frac{\text{work output}}{\text{heat input}} \qquad \text{or} \qquad e = \frac{P\,\Delta r \cos\theta}{|Q_h|}$$

So the distance the ship can travel per kilogram of uranium fuel is

$$\Delta r = \frac{e|Q_h|}{P \cos 0°} = \frac{0.200(2.90 \times 10^{12} \text{ J})}{1.00 \times 10^5 \text{ N}} = 5.80 \times 10^6 \text{ m} \qquad \Diamond$$

The ship can travel 5800 km/kg of uranium fuel, which is on the high end of our prediction range. The distance between New York and Paris is 5851 km, so the ship could cross the Atlantic ocean with just one kilogram of uranium fuel.

---

35. It has been estimated that about 10^9 tons of natural uranium is available at concentrations exceeding 100 parts per million, of which 0.7% is the fissionable isotope ^{235}U. Assume that all the world's energy use $(7 \times 10^{12}$ J/s) were supplied by ^{235}U fission in conventional nuclear reactors, releasing 208 MeV for each reaction. How long would the supply last? The estimate of uranium supply is taken from K. S. Deffeyes and I. D. MacGregor, "World Uranium Resources." *Sci. Am.*, **242(1)**: 66, 1980.

Solution   The mass of natural uranium reserves is

$$m_{U,\ total} = \left(10^9 \text{ metric tons}\right)\left(\frac{10^3 \text{ kg}}{1 \text{ metric ton}}\right) = 1 \times 10^{12} \text{ kg}$$

For fissionable U-235, $m_{U\text{-}235} = (0.007)\left(m_{U,\ total}\right) = 7 \times 10^9 \text{ kg} = 7 \times 10^{12} \text{ g}$

In terms of U-235 nuclei, $N_{235} = \left(7 \times 10^{12} \text{ g}\right)\left(\dfrac{6.02 \times 10^{23} \text{ atoms}}{235 \text{ g}}\right) = 1.79 \times 10^{34} \text{ nuclei}$

Following the statement of the problem, we take the fission energy as 208 MeV/fission:

$$E = \left(1.79 \times 10^{34} \text{ nuclei}\right)(208 \text{ MeV/fission})\left(1.60 \times 10^{-19} \text{ J/eV}\right) = 5.97 \times 10^{23} \text{ J}$$

Now, $\Delta t = \dfrac{E}{\mathcal{P}} = \dfrac{5.97 \times 10^{23} \text{ J}}{7 \times 10^{12} \text{ J/s}} = 8.53 \times 10^{10} \text{ s} \sim 3000 \text{ yr}$ ◊

---

37. Consider the two nuclear reactions

(I) $A + B \rightarrow C + E$

and   (II) $C + D \rightarrow F + G$

(a) Show that the net disintegration energy for these two reactions $(Q_{net} = Q_I + Q_{II})$ is identical to the disintegration energy for the net reaction

$$A + B + D \rightarrow E + F + G$$

(b) One possible chain of reactions in the proton-proton cycle in the Sun's core is

$^1_1H + {}^1_1H \rightarrow {}^2_1H + e^+ + \nu$ $e^+ + e^- \rightarrow 2\gamma$ $^1_1H + {}^2_1He \rightarrow {}^3_2He + \gamma$

$^1_1H + {}^3_2He \rightarrow {}^4_2He + e^+ + \nu$ $e^+ + e^- \rightarrow 2\gamma$

Based on part (a), what is Q_{net} for this sequence?

Solution

$$Q_I = (M_A + M_B - M_C - M_D)c^2$$

$$Q_{II} = (M_C + M_D - M_F - M_G)c^2$$

$$Q_{net} = (M_A + M_B - M_C - M_E + M_C + M_D - M_F - M_G)c^2$$

$$Q_{net} = (M_A + M_B + M_D - M_E - M_F - M_G)c^2$$

(a) This value is identical to Q for the reaction $A + B + D \rightarrow E + F + G$. Thus, any product (e.g., "C") that is a reactant in a subsequent reaction disappears from the energy balance. ◊

(b) Adding all five reactions,

we have $\qquad$ $4\left(^1_1H\right) + 2\left(^0_{-1}e\right) \rightarrow ^4_2He + 2\nu$

Here the symbol ^1_1H represents a proton, the nucleus of a hydrogen-1 atom. So that we may use the tabulated masses of neutral atoms for calculation, we add two electrons to each side of the reaction. The four electrons on the initial side are enough to make four hydrogen atoms and the two electrons on the final side constitute, with the alpha particle, a neutral helium atom. The atomic-electronic binding energies are negligible compared to Q_{net}.

We use the arbitrary symbol "^1_1H atom" to represent a neutral atom.

Then we have $\qquad$ $4\left(^1_1H \text{ atom}\right) \rightarrow ^4_2He \text{ atom} + 2\nu$

$$4(1.007825 \text{ u}) = 4.002602 \text{ u} + \frac{Q_{net}}{c^2}$$

$$Q_{net} = [4(1.007\,825 \text{ u}) - 4.002\,602 \text{ u}](931.5 \text{ MeV/u}) = 26.7 \text{ MeV} \quad ◊$$

49. The decay of an unstable nucleus by alpha emission is represented by Equation 30.9. The disintegration energy Q given by Equation 30.12 must be shared by the alpha particle and the daughter nucleus to conserve both energy and momentum in the decay process. (a) Show that Q and K_α, the kinetic energy of the alpha particle, are related by the expression

$$Q = K_\alpha\left(1 + \frac{M_\alpha}{M}\right)$$

where M is the mass of the daughter nucleus. (b) Use the result of part (a) to find the energy of the alpha particle emitted in the decay of ^{226}Ra. (See Example 30.5 for the calculation of Q.)

Solution   For clarity, we will let the subscript d denote the daughter nucleus.

(a)   Let us assume that the parent nucleus (mass M_p) is initially at rest. We denote the masses of the daughter nucleus and alpha particle by M_d and M_α, respectively. Applying the equations of conservation of momentum and energy for the alpha decay process

gives

$$M_d v_d = M_\alpha v_\alpha \tag{1}$$

$$M_p c^2 = M_d c^2 + M_\alpha c^2 + \frac{1}{2}M_\alpha v_\alpha{}^2 + \frac{1}{2}M_d v_d{}^2 \tag{2}$$

The disintegration energy Q is given by

$$Q = (M_p - M_d - M_\alpha)c^2 = \frac{1}{2}M_\alpha v_\alpha{}^2 + \frac{1}{2}M_d v_d{}^2 \tag{3}$$

Eliminating v_d from Equations (1) and (3)

gives

$$Q = \frac{1}{2}M_\alpha v_\alpha{}^2 + \frac{1}{2}M_d\left(\frac{M_\alpha}{M_d}v_\alpha\right)^2 = \frac{1}{2}M_\alpha v_\alpha{}^2 + \frac{1}{2}\frac{M_\alpha{}^2}{M_d}v_\alpha{}^2$$

$$Q = \frac{1}{2}M_\alpha v_\alpha{}^2\left(1 + \frac{M_\alpha}{M_d}\right) = K_\alpha\left(1 + \frac{M_\alpha}{M_d}\right) \qquad \lozenge$$

(b)

$$K_\alpha = \frac{Q}{1 + \dfrac{M_\alpha}{M_d}} = \frac{4.87 \text{ MeV}}{1 + \dfrac{4}{222}} = 4.78 \text{ MeV} \qquad \lozenge$$

Chapter 31

Particle Physics and Cosmology

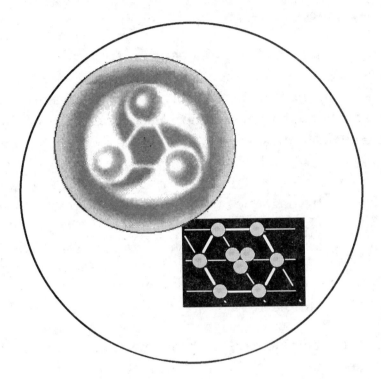

INTRODUCTION

In this chapter, we examine the properties and classifications of the various known subatomic particles and the fundamental interactions that govern their behavior. We also discuss the current theory of elementary particles, in which all matter is believed to be constructed from only two families of particles, quarks and leptons. Finally, we discuss how clarifications of such models might help scientists understand cosmology, which deals with the evolution of the Universe.

NOTES FROM SELECTED CHAPTER SECTIONS

31.1   The Fundamental Forces in Nature

There are four fundamental forces in nature: **strong** (hadronic), **electromagnetic**, **weak**, and **gravitational**. A residual effect of the strong force is the nuclear force, which acts between nucleons to keep the nucleus together. The weak force is responsible for beta decay. The electromagnetic and weak forces are now considered to be manifestations of a single force called the **electroweak** force.

The fundamental forces are described in terms of particle or quanta exchanges which **mediate** the forces. The electromagnetic force is mediated by photons, which are the quanta of the electromagnetic field. Likewise, the strong force is fundamentally mediated by field particles called **gluons**, the weak force is mediated by particles called the W and Z **bosons**, and the gravitational force is mediated by quanta of the gravitational field called **gravitons**.

31.2   Positrons and Other Antiparticles

An antiparticle and a particle have the same mass, but opposite charge. Furthermore, other properties may have opposite values such as lepton number and baryon number. It is possible to produce particle-antiparticle pairs in nuclear reactions if the available energy is greater than $2mc^2$, where m is the mass of the particle.

Pair production is a process in which a gamma ray with an energy of at least 1.02 MeV interacts with a nucleus, and an electron-positron pair is created.

Pair annihilation is an event in which an electron and a positron can annihilate to produce two gamma rays, each with an energy of at least 0.511 MeV.

31.3   Mesons and the Beginning of Particle Physics

The interaction between two particles can be represented in a diagram called a **Feynman diagram**.

31.4   Classification of Particles

All particles (other than field particles) can be classified into two categories: **hadrons** and **leptons**.

There are two classes of hadrons: **mesons** and **baryons**, which are grouped according to their masses and spins. It is believed that hadrons are composed of units called **quarks**, which are fundamental in nature.

Leptons have no structure or size and are therefore considered to be truly elementary particles.

31.5   Conservation Laws

In all reactions and decays, quantities such as energy, linear momentum, angular momentum, electric charge, baryon number, and lepton number are strictly conserved. Certain particles have properties called **strangeness** and **charm**. These unusual properties are conserved only in those reactions and decays that occur via the strong and electromagnetic forces.

Whenever a nuclear reaction or decay occurs, the sum of the baryon numbers before the process must equal the sum of the baryon numbers after the process.

The sum of the electron-lepton numbers before a reaction or decay must equal the sum of the electron-lepton numbers after the reaction or decay.

31.6   Strange Particles and Strangeness

Whenever the strong force or the electromagnetic force causes a reaction or decay to occur, the sum of the strangeness numbers before the process must equal the sum of the strangeness numbers after the process.

31.9   Quarks

Recent theories in elementary particle physics have postulated that all hadrons are composed of smaller units known as **quarks**. Quarks have a fractional electric charge and a baryon number of 1/3. There are six flavors of quarks, up (u), down (d), strange (s), charmed (c), top (t), and bottom (b). All baryons contain three quarks, while all mesons contain one quark and one antiquark.

According to the theory of **quantum chromodynamics**, quarks have a property called **color**, and the strong force between quarks is referred to as the **color force**.

EQUATIONS AND CONCEPTS

Pions and muons are very unstable particles. A decay sequence is shown in Equation 31.1.

$$\pi^- \to \mu^- + \bar{\nu}^-$$ (31.1)

$$\mu^- \to e^- + \nu + \bar{\nu}^-$$

Hubble's law states a linear relationship between the velocity of a galaxy and its distance R from the Earth. The constant H is called the **Hubble parameter**.

$$v = HR$$ (31.7)

$$H = 17 \times 10^{-3} \text{ m/s} \cdot \text{light-year}$$

Each baryon and meson is composed of quarks and antiquarks; the identity of each particle is based upon the combination of quarks involved. When a particle decays, a quark can annihilate with its corresponding antiquark; alternatively, a quark-antiquark pair can be spontaneously formed. However, in any interaction, the baryon number is always conserved.

Quark, Antiquark	Charge	Baryon Number
u, $\bar{u}$	$+\frac{2}{3}e, \; -\frac{2}{3}e$	$\frac{1}{3}, \; -\frac{1}{3}$
d, $\bar{d}$	$-\frac{1}{3}e, \; +\frac{1}{3}e$	$\frac{1}{3}, \; -\frac{1}{3}$
s, $\bar{s}$	$-\frac{1}{3}e, \; +\frac{1}{3}e$	$\frac{1}{3}, \; -\frac{1}{3}$
c, $\bar{c}$	$+\frac{2}{3}e, \; -\frac{2}{3}e$	$\frac{1}{3}, \; -\frac{1}{3}$
b, $\bar{b}$	$-\frac{1}{3}e, \; +\frac{1}{3}e$	$\frac{1}{3}, \; -\frac{1}{3}$
t, $\bar{t}$	$+\frac{2}{3}e, \; -\frac{2}{3}e$	$\frac{1}{3}, \; -\frac{1}{3}$

REVIEW CHECKLIST

▷ Be aware of the four fundamental forces in nature and the corresponding field particles or quanta via which these forces are mediated.

▷ Understand the concepts of the antiparticle, pair production, and pair annihilation.

▷ Know the broad classification of particles and the characteristic properties of the several classes (relative mass value, spin, decay mode).

ANSWERS TO SELECTED CONCEPTUAL QUESTIONS

4. Describe the properties of baryons and mesons and the important differences between them.

Answer There are two types of hadrons, called baryons and mesons. Hadrons interact primarily through the strong force and are not elementary particles, being composed of either three quarks (baryons), or a quark and an antiquark (mesons). Baryons have a nonzero baryon number with a spin of either 1/2 or 3/2. Mesons have a baryon number of zero, and a spin of either 0 or 1.

7. The Ξ^0 particle decays by the weak interaction according to the decay mode $\Xi^0 \rightarrow \Lambda^0 + \pi^0$. Would you expect this decay to be fast or slow? Explain.

Answer This decay should be slow, since decays which occur via the weak interaction typically take 10^{-10} s or longer to occur.

9. Identify the particle decays listed in Table 31.2 that occur by the electromagnetic interaction. Justify your answers.

Answer The decays of the neutral pion, eta, and neutral sigma occur by the electromagnetic interaction. These are the three shortest lifetimes in Table 31.2. All produce photons, which are the quanta of the electromagnetic force. All conserve strangeness.

14. How many quarks are there in (a) a baryon, (b) an antibaryon, (c) a meson, (d) an antimeson? How do you account for the fact that baryons have half-integral spins and mesons have spins of 0 or 1? (**Hint:** Quarks have spin 1/2.)

Answer All baryons and antibaryons consist of three quarks. All mesons and antimesons consist of two quarks. Since quarks have spins of 1/2, it follows that all baryons (which consist of three quarks) must have half-integral spins, and all mesons (which consist of two quarks) must have spins of 0 or 1.

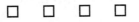

SOLUTIONS TO SELECTED END-OF-CHAPTER PROBLEMS

3. A photon with an energy $E_\gamma = 2.09$ GeV creates a proton-antiproton pair in which the proton has a kinetic energy of 95.0 MeV. What is the kinetic energy of the antiproton ($m_p c^2 = 938.3$ MeV)?

Solution The total energy of each particle is the sum of its rest energy and its kinetic energy. To describe the creation process, we choose as our system the photon before the reaction and the proton and antiproton together afterward. Conservation of energy in the creation process requires that the total energy before this pair production event equal the total energy after.

$$E_\gamma = \left(E_{Rp} + K_p\right) + \left(E_{R\bar{p}} + K_{\bar{p}}\right)$$

The energy of the photon is given as $E_\gamma = 2.09$ GeV $= 2.09 \times 10^3$ MeV. From Table 31.2, we see that the rest energy of both the proton and the antiproton is

$$E_{Rp} = E_{R\bar{p}} = m_p c^2 = 938.3 \text{ MeV}$$

If the kinetic energy of the proton is observed to be 95.0 MeV, the kinetic energy of the antiproton is

$$K_{\bar{p}} = E_\gamma - E_{R\bar{p}} - E_{Rp} - K_p = 2.09 \times 10^3 \text{ MeV} - 2(938.3 \text{ MeV}) - 95.0 \text{ MeV} = 118 \text{ MeV} \qquad \lozenge$$

---

5. One of the mediators of the weak interaction is the Z^0 boson, with mass 93 GeV/c^2. Use this information to find the order of magnitude of the range of the weak interaction.

Solution We use the quantum particle model. The rest energy of the Z^0 boson is $E_0 = 93$ GeV. The maximum time a virtual Z^0 boson can exist is found from

$$\Delta E \, \Delta t \geq \frac{1}{2}\hbar \qquad \text{or} \qquad \Delta t \cong \frac{\hbar}{2\,\Delta E} = \frac{1.055 \times 10^{-34} \text{ J}\cdot\text{s}}{2(93 \text{ GeV})\left(1.60 \times 10^{-10} \text{ J/GeV}\right)} = 3.55 \times 10^{-27} \text{ s}$$

The maximum distance it can travel in this time is

$$d = c\,\Delta t = \left(3.00 \times 10^8 \text{ m/s}\right)\left(3.55 \times 10^{-27} \text{ s}\right) \sim 10^{-18} \text{ m} \qquad \lozenge$$

The distance d is an approximate value for the range of the weak interaction.

9. A neutral pion at rest decays into two photons according to $\pi^0 \rightarrow \gamma + \gamma$.

Find the energy, momentum, and frequency of each photon.

Solution We use the energy and momentum versions of the isolated system model.

Since the pion is at rest and momentum is conserved in the decay, the two gamma-rays must have equal amounts of momentum in opposite directions. So, they must share equally in the energy of the pion.

$$m_{\pi^0} = 135.0 \text{ MeV}/c^2 \text{ (Table 31.2)}$$

Therefore, $E_\gamma = 67.5 \text{ MeV} = 1.08 \times 10^{-11} \text{ J}$ ◊

$$p = \frac{E_\gamma}{c} = \frac{67.5 \text{ MeV}}{3.00 \times 10^8 \text{ m/s}} = 3.61 \times 10^{-20} \text{ kg} \cdot \text{m/s}$$ ◊

$$f = \frac{E_\gamma}{h} = 1.63 \times 10^{22} \text{ Hz}$$ ◊

11. Name one possible decay mode (see Table 31.2) for Ω^+, $\overline{K}_S^0$, $\overline{\Lambda}^0$, and $\overline{n}$.

Solution The particles in this problem are the antiparticles of those listed in Table 31.2. Therefore, the decay modes include the antiparticles of those shown in the decay modes in Table 31.2:

$$\Omega^+ \rightarrow \overline{\Lambda}^0 + K^+$$

$$\overline{K}_S^0 \rightarrow \pi^+ + \pi^- \qquad \text{or} \qquad \pi^0 + \pi^0$$

$$\overline{\Lambda}^0 \rightarrow \overline{p} + \pi^+$$

$$\overline{n} \rightarrow \overline{p} + e^+ + \nu_e$$ ◊

15. The following reactions or decays involve one or more neutrinos. In each case supply the missing neutrino (v_e, v_μ, or $_\tau$).

(a) $\pi^- \to \mu^- + ?$ (b) $K^+ \to \mu^+ + ?$ (c) $? + p \to n + e^+$

(d) $? + n \to p + e^-$ (e) $? + n \to p + \mu^-$ (f) $\mu^- \to e^- + ? + ?$

Solution

(a) $\pi^- \to \mu^- + \bar{v}_\mu$ $L_\mu:\ 0 \to 1 - 1$ ◊

(b) $K^+ \to \mu^+ + v_\mu$ $L_\mu:\ 0 \to -1 + 1$ ◊

(c) $\bar{v}_e + p^+ \to n + e^+$ $L_e:\ -1 + 0 \to 0 - 1$ ◊

(d) $v_e + n \to p^+ + e^-$ $L_e:\ 1 + 0 \to 0 + 1$ ◊

(e) $v_\mu + n \to p^+ + \mu^-$ $L_\mu:\ 1 + 0 \to 0 + 1$ ◊

(f) $\mu^- \to e^- + \bar{v}_e + v_\mu$ $L_\mu:\ 1 \to 0 + 0 + 1$ and $L_e:\ 0 \to 1 - 1 + 0$ ◊

---

17. Determine which of the reactions can occur. For those that cannot, determine the conservation law (or laws) violated.

(a) $p \to \pi^+ + \pi^0$ (b) $p + p \to p + p + \pi^0$ (c) $p + p \to p + \pi^+$

(d) $\pi^+ \to \mu^+ + v_\mu$ (e) $n \to p + e^- + \bar{v}_e$ (f) $\pi^+ \to \mu^+ + n$

Solution

(a) $p \to \pi^+ + \pi^0$ Baryon number is violated: $1 \to 0 + 0$ ◊

(b) $p + p \to p + p + \pi^0$ This reaction can occur.   ◊

(c) $p + p \to p + \pi^+$ Baryon number is violated: $1 + 1 \to 1 + 0$ ◊

(d) $\pi^+ \to \mu^+ + v_\mu$ This reaction can occur.   ◊

(e) $n \to p + e^- + \bar{v}_e$ This reaction can occur.   ◊

(f) $\pi^+ \to \mu^+ + n$ Violates baryon number: $0 \to 0 + 1$

     and   Violates muon-lepton number: $0 \to -1 + 0$ ◊

---

21. Determine whether or not strangeness is conserved in the following decays and reactions:

(a) $\Lambda^0 \to p + \pi^-$ (b) $\pi^- + p \to \Lambda^0 + K^0$ (c) $\bar{p} + p \to \bar{\Lambda}^0 + \Lambda^0$

(d) $\pi^- + p \to \pi^- + \Sigma^+$ (e) $\Xi^- \to \Lambda^0 + \pi^-$ (f) $\Xi^0 \to p + \pi^-$

Solution   We look up the strangeness quantum numbers in Table 31.2.

(a) $\Lambda^0 \to p + \pi^-$ Strangeness: $-1 \to 0 + 0$

(-1 does not equal 0, so strangeness is not conserved)   ◊

(b) $\pi^- + p \to \Lambda^0 + K^0$ Strangeness: $0 + 0 \to -1 + 1$

($0 = 0$ and strangeness is conserved)   ◊

(c) $\bar{p} + p \to \bar{\Lambda}^0 + \Lambda^0$ Strangeness: $0 + 0 \to +1 - 1$

($0 = 0$ and strangeness is conserved)   ◊

(d) $\pi^- + p \to \pi^- + \Sigma^+$ Strangeness: $0 + 0 \to 0 - 1$

(0 does not equal -1 so strangeness is not conserved)   ◊

(e) $\Xi^- \to \Lambda^0 + \pi^-$ Strangeness: $-2 \to -1 + 0$

(-2 does not equal -1 so strangeness is not conserved)   ◊

(f) $\Xi^0 \to p + \pi^-$ Strangeness: $-2 \to 0 + 0$

(-2 does not equal 0 so strangeness is not conserved)   ◊

---

27. If a K_S^0 meson at rest decays in 0.900×10^{-10} s, how far will a K_S^0 meson travel if it is moving at $0.960c$ through a bubble chamber?

Solution   The motion of the K_S^0 particle is relativistic. Just like the spaceman who leaves for a distant star, and returns to find his family long gone, the kaon appears to us to have a longer lifetime. That time-dilated lifetime is:

$$t = \gamma t_0 = \frac{0.900 \times 10^{-10} \text{ s}}{\sqrt{1 - v^2/c^2}} = \frac{0.900 \times 10^{-10} \text{ s}}{\sqrt{1 - (0.960)^2}} = 3.21 \times 10^{-10} \text{ s}$$

During this time, we see the kaon travel at $0.960c$. It travels for a distance of

$$d = vt = (0.960)(3.00 \times 10^8 \text{ m/s})(3.21 \times 10^{-10} \text{ s}) = 0.0926 \text{ m} = 9.26 \text{ cm}$$ ◊

---

33. Analyze each reaction in terms of constituent quarks:

(a) $\pi^- + p \rightarrow K^0 + \Lambda^0$

(b) $\pi^+ + p \rightarrow K^+ + \Sigma^+$

(c) $K^- + p \rightarrow K^+ + K^0 + \Omega^-$

(d) $p + p \rightarrow K^0 + p + \pi^+ + ?$

In the last reaction, identify the mystery particle.

Solution We look up the quark constituents of the particles in Table 31.4.

(a) $d\bar{u} + uud \rightarrow d\bar{s} + uds$ ◊

(b) $\bar{d}u + uud \rightarrow u\bar{s} + uus$ ◊

(c) $\bar{u}s + uud \rightarrow u\bar{s} + d\bar{s} + sss$ ◊

(d) $uud + uud \rightarrow d\bar{s} + uud + u\bar{d} + uds$ (A uds is either a Λ^0 or a Σ^0) ◊

---

47. The energy flux carried by neutrinos from the Sun is estimated to be on the order of $0.4 \ \text{W/m}^2$ at Earth's surface. Estimate the fractional mass loss of the Sun over 10^9 years due to the emission of neutrinos. (The mass of the Sun is $2 \times 10^{30} \ \text{kg}$. The Earth-Sun distance is $1.5 \times 10^{11} \ \text{m}$.)

Solution Since the neutrino flux from the Sun reaching the Earth is $0.4 \ \text{W/m}^2$, the total energy emitted per second by the Sun in neutrinos is

$$\left(0.4 \ \text{W/m}^2\right)\left(4\pi r^2\right) = \left(0.4 \ \text{W/m}^2\right)\left[4\pi\left(1.5 \times 10^{11} \ \text{m}\right)^2\right] = 1.13 \times 10^{23} \ \text{W}$$

In a period of 10^9 years, the Sun emits a total energy of

$$\left(1.13 \times 10^{23} \ \text{J/s}\right)\left(10^9 \ \text{yr}\right)\left(3.156 \times 10^7 \ \text{s/yr}\right) = 3.56 \times 10^{39} \ \text{J}$$

in the form of neutrinos. This energy corresponds to an annihilated mass of

$$mc^2 = 3.56 \times 10^{39} \ \text{J} \qquad m = \frac{E}{c^2} = \frac{3.56 \times 10^{39} \ \text{J}}{\left(3.00 \times 10^8 \ \text{m/s}\right)^2} = 3.96 \times 10^{22} \ \text{kg}$$

Since the Sun has a mass of about $2 \times 10^{30} \ \text{kg}$, this corresponds to a loss of only about 1 part in $50\,000\,000$ of the Sun's mass over 10^9 years in the form of neutrinos. ◊

---

49. Free neutrons have a characteristic half-life of 10.4 min. What fraction of a group of free thermal neutrons with kinetic energy 0.0400 eV will decay before traveling a distance of 10.0 km?

Solution

The fraction that will remain is given by the ratio N/N_0,

where $$N/N_0 = e^{-\lambda \Delta t}$$

and Δt is the time it takes the neutron to travel a distance of $d = 10.0$ km.

We use the particle under constant speed model. The time of flight is given by $\Delta t = d/v$.

Since $K = \frac{1}{2}mv^2$, $$\Delta t = \frac{d}{\sqrt{\dfrac{2K}{m}}} = \frac{10.0 \times 10^3 \text{ m}}{\sqrt{\dfrac{2(0.0400 \text{ eV})(1.60 \times 10^{-19} \text{ J/eV})}{1.67 \times 10^{-27} \text{ kg}}}} = 3.61 \text{ s}$$

The decay constant is then $$\lambda = \frac{0.693}{T_{1/2}}$$

$$\lambda = \frac{0.693}{(10.4 \text{ min})(60 \text{ s/min})} = 1.11 \times 10^{-3} \text{ s}^{-1}$$

Therefore, $$\lambda \Delta t = \left(1.11 \times 10^{-3} \text{ s}^{-1}\right)(3.61 \text{ s})$$

$$\lambda \Delta t = 4.01 \times 10^{-3} = 0.00401$$

and $$\frac{N}{N_0} = e^{-\lambda \Delta t} = e^{-0.00401} = 0.9960$$

Hence, the fraction that has decayed in this time is

$$1 - \frac{N}{N_0} = 0.00400 \quad \text{or} \quad 0.400\% \qquad \Diamond$$

51. Determine the kinetic energies of the proton and pion resulting from the decay of a Λ^0 at rest:

$$\Lambda^0 \rightarrow p + \pi^-$$

Solution   We first look up the energy of each particle:

$$m_\Lambda c^2 = 1115.6 \text{ MeV}$$

$$m_p c^2 = 938.3 \text{ MeV}$$

$$m_\pi c^2 = 139.6 \text{ MeV}$$

The difference between starting mass-energy and final mass-energy is the kinetic energy of the products:

$$K_p + K_\pi = (1115.6 - 938.3 - 139.6) \text{ MeV} = 37.7 \text{ MeV}$$

In addition, since momentum is conserved in the decay,

$$\left|p_p\right| = \left|p_\pi\right| = p$$

We use one symbol for the magnitude of the momentum of each product particle.

Applying conservation of relativistic energy:

$$\left(\sqrt{(938.3)^2 + p^2 c^2} - 938.3\right) + \left(\sqrt{(139.6)^2 + p^2 c^2} - 139.6\right) = 37.7 \text{ MeV}$$

Solving the algebra yields $p_\pi c = p_p c = 100.4 \text{ MeV}$

Thus,

$$K_p = \sqrt{\left(m_p c^2\right)^2 + (100.4)^2} - m_p c^2 = 5.35 \text{ MeV} \qquad \Diamond$$

and

$$K_\pi = \sqrt{(139.6)^2 + (100.4)^2} - 139.6 = 32.3 \text{ MeV} \qquad \Diamond$$